REPORT

OF

BOTANICAL SURVEY

OF

SOUTHERN AND CENTRAL LOUISIANA,

MADE DURING THE YEAR 1870,

By A. FEATHERMAN,

LECTURER ON BOTANY AND PROFESSOR LOUISIANA STATE UNIVERSITY.

NEW ORLEANS:
PRINTED AT THE OFFICE OF THE REPUBLICAN, 94 CAMP STREET.

1871.

REPORT

OF

Botanical Survey of Southern and Central Louisiana,

MADE DURING THE YEAR 1870,

By A. FEATHERMAN,

Lecturer on Botany and Professor Louisiana State University.

LOUISIANA STATE UNIVERSITY, January 7, 1871.

Colonel D. F. Boyd, Superintendent Louisiana State University:

THE MORAL AND INTELLECTUAL INFLUENCE OF BOTANICAL STUDIES.

The botanical survey of the State of Louisiana is not only important in a scientific point of view, but it can not fail to be productive of beneficial practical results. Botany, as a branch of the natural sciences, produces a more salutary influence to elevate and refine society, and raise the moral standard of civilization, than all the other sciences combined. Flowers are familiar friends and loved companions in every household, where refinement and virtue give tone and character to the social circle. I can do no better than quote, in this connection, the eloquent words of an able English writer, and apply them to the study of botany in particular: "It is fearfully true that nine-tenths of the immorality which pervades the better classes of society, originate from the want of an interesting occupation to fill up the vacant time; and as the study of botany is as attractive as it is beneficial, it must necessarily exert a moral and even religious influence upon the young and inquiring mind.

"The youth, who is fond of scientific pursuits, will not enter into revelry, for frivolous and vicious excitement will have no fascination for him. The overflowing cup, the unmeaning or dishonest game will not excite him. If any one doubts the beneficial influence of these studies on the morals and character, I would ask him to point out the immoral young man who is devotedly attached to any branch of natural science. I never knew such a one. There may be such individuals—for religion only can change the heart—but if there be they are very rare exceptions, and the loud clamors which are always raised against the man of science who errs proves how rarely the study of the Creator fails to exert an ennobling effect upon a well regulated mind. Fortunate, indeed, are the youths of either sex who early imbibe a taste for naturàl knowledge, and whose predilections are not thwarted by injudicious friends."

It may indeed be said of the botanist what has been said of the astronomer, that a botanist who does not believe in the existence of God is mad, for the sublime beauties of nature which are constantly presented to his mind, must necessarily lead him from nature to nature's God, whom he recognizes as the author of all those splendid marvels of creative wisdom, which are scattered all around him for the use and enjoyment of man.

THE IMPORTANCE OF BOTANY IN AGRICULTURE AND MEDICINE.

Botany has an intimate connection with agriculture, and its practical bearing in this respect has not been sufficiently developed. The agriculturalist who combines science with experience and practical observation, may derive some valuable hints from his botanical knowledge, which might lead to the most important results, not only in the cultivation of the plants, which are the sources of his wealth, but in banishing from his fields those troublesome weeds and grasses which keep his plows and hoes busy during the whole period of the crop season. It is indeed a very remarkable fact that even physicians, otherwise well educated, know little or nothing about botanical science, when more than one-half of all the remedial agents employed, and some of them the most important of the pharmacopœia, are derived from the vegetable kingdom, while the other half, comprising mineral substances altogether incompatible with human organism, might be beneficially superseded by vegetable remedies of far greater efficacy and active powers, if physicians studied the numerous plants possessing medicinal virtues, which are diffused in the greatest abundance all over the surface of the globe. None of our medical schools has a chair of botany, which is certainly as great an anomaly in the educational progress of a nation as if a law school had no professorship expounding the principles of the common and civil laws, which must be the foundation of the lawyer's professional acquirements, or if a theological school failed to require a knowledge of the Greek and Hebrew languages, which alone can enable the candidate for orders to expound the scriptures in spirit and in truth.

Besides your other valuable services in behalf of education the State of Louisiana owes you a debt of gratitude for the exertions made by you in providing the means to prosecute the botanical survey of the State, which will acquire for the University—strictly a State institution—one of the most valuable scientific collections, that will last for centuries if properly cared for, and which money could not buy nor favor procure. It is fit that Louisiana, the Empire State of the South-west, with its immense agricultural resources and its network of navigable streams, should be the first of all the States to place botany side by side with geology in the investigation of the physical and economic resources of the State.

The popular idea of a geological survey is that the geologist is prospecting—using an expressive term of the California miner—in order to discover the Ophir where gold is as abundant as common dust; or to find the New Castle where coal constitutes the solid crust of the earth. But these popular notions about geology have no application to Louisiana, for the nature of the geological formation of the State, is sufficient for the scientific geologist to determine at once, that these materials do not and can not exist here in a form to give them economic value. But while geology has undoubtedly a scientific and a practical side, botany has no less so. It is a curious fact in the history of education, that educated men should be entirely ignorant of the vegetable kingdom, which furnishes us with the bread we eat, the wine we drink, the fruits we relish, the clothing we wear, the most ornamental parts of the dwellings we inhabit, the light which changes the night into day, the ships as well as the materials of commerce which they transport, the fuel which equalizes the seasons and supplies the motive power of the steam engine and the locomotive. The vegetable kingdom, without which no living creature that breathes can live, is the most rich in resources and the most important in a commercial, artistic, mechanical and agricultural point of view. The study of botany is not only of the greatest interest to the natural philosopher, but it deserves to be taught in every school, on account of its attractive features and the educational discipline it affords. It occupies, morally and mentally, a far higher rank than chemistry or geology, which treat only of inert matter and are acted upon by external forces, while botany treats of. living organisms of complicated structure and wonderful arrangement.

SCIENTIFIC EDUCATION OF THE PLANTER AND FARMER.

Considering that the agriculturalist derives his wealth exclusively from the vegetable kingdom, and is, so to speak, the lord and master, who monopolizes its most valuable products, forcing nature by his constant attention and unremitting industry to yield up a hundred-fold the treasures confided to the fertile soil, it is still more strange that the educated planter and farmer should never have bestowed a moment's reflection upon the marvelous productiveness of the seed he sows broadcast over his land, and should never have investigated the principle of its germination, growth and development, and examined the structure and the nutritive element of the plants which are the life of his life and the source of his prosperity. He makes his daily rounds in his fields, watches the progressive advancement of his crops, counts up in his mind what quantity of sugar, cotton, corn, rice, wheat, oats, tobacco, or indigo each acre may yield, considers its actual value in money, but it never occurs to his mind that all these are living, self-supporting, and self-producing organisms, which God has created not merely for utilitarian purposes to supply the wants of the animal world, but as manifestations of His goodness and His wisdom, as living monitors to teach man his duty to labor and render himself useful, and to inspire him with an aim higher and nobler than merely to eat and drink and get rich, but to disenthral his soul from the enslaving materialistic ideas which cling to him and make him a mean and groveling creature.

The planters and farmers constitute in every country the majority of the population, and they are the most useful class of society. But by a perversion of human reason they have obtained but little attention from grovernments, except as a tax-paying class. We have, however, reason to be proud of our country, for its government has recognized the fact that the stability of free institutions must rest upon an educated and refined yeomanry, and it has consequently granted munificent donations to the States for the establishment of agricultural schools, where the planter and farmer may not only receive a literary but a scientific and professional education. Medical schools, law schools and theological schools have been in existence for centuries, but a school for agriculture, the most useful and the most wide-spread of all professions, would have been considered, even fifty years ago, the most absurd of all institutions.

The agricultural college of Louisiana, which is soon to be established, should form one of the departments of the University, for it offers all the advantages such a school requires without much additional labor and expense. It has the geological and mineralogical cabinet. It will have one of the most extensive botanical museums of the South. It has already professors of chemistry, geology, mathematics, engineering, botany, modern languages, and it has also a commercial department, which are the principal branches taught in agricultural schools. It only needs in addition professors of applied chemistry, of practical agriculture and the mechanic arts. As the funds derived from the sale of the land scrip can not be diverted for the construction of college buildings, the University buildings, wheresoever located, could be so constructed as to combine both schools, and would thus save the State considerable expense.

AGRICULTURAL RESOURCES OF LOUISIANA.

The agricultural resources of Louisiana are very great. There is no State in the Union, and no country on the globe of equal dimensions, which can be compared with her in the salubrity and mildness of her climate; the extent of her alluvial and arable lands; the variety of her agricultural industry; the numerous navigable streams, by which she is intersected, and her facilities for ocean navigation. A State for which nature has done so much, and man has done so little; a State where the orange and banana ripen their fruit in the open air, and give to it an intertropical character; a State where cotton, the most valuable product that feeds the commerce of the world and regulates international finances, flourishes in its highest perfection; where sugar, the indispensable luxury of every household, can be produced in quantities sufficient to supply forty per cent. of all that is consumed in the United States; where rice is cultivated for exportation, where corn grows most luxuriantly, where tobacco of superior quality can be produced, and wheat and oats, and all other cereals and almost every other product of the field, the garden and the orchard would thrive with a vigor unsurpassed by any other climate; a State that possesses such invaluable natural advantages, should be foremost in science and literature, and in institutions, created and designed for the diffusion of useful knowledge. It should strive to become the modern Greece of America for learning and refinement; and the Italy of the western continent for its inexhaustible fertility, and the neatness, beauty and high culture of its farms and plantations.

I trust I may be pardoned for this ardor in the cause of civilization, but I feel a deep interest in the State of Louisiana. I know her people to be a peculiar people; they, like the Virginians, still retain some State pride, they never call themselves by the indefinite and vague name of Americans, but they are Louisianians in manners, sentiments and feelings. They combine the disinterestedness and humane sympathies of the French and Spaniard with the chivalric bearing and the practical sense of the Englishman. I know what civilization ought to expect of them; and I equally know that they can and will accomplish the high destiny reserved for them, which will fill a noble page in the world's history.

SOIL OF LOUISIANA, DIVIDED INTO CLASSES.

The alluvial lands of Louisiana, protected by an effectual levee system, would present the most magnificent agricultural domain that can be found anywhere within the limits of the same extent of country.

The object which I had in view in traveling through the State enabled me only to visit about eight or nine parishes. But I was struck with astonishment at the inexhaustless fertility of the soil extending over such a vast area, as I was passing in my excursions through the alluvial and marsh prairie regions of the Red river, the Mississippi, Bayou Barataria, the Atchafalaya and the Teche, comprising millions of acres of the most fertile and most productive lands. The alluvial lands below Red river are estimated at 7,860,000 acres, which, if cultivated, would be capable of supporting a population of ten millions of people, in addition to the fertile lands in the interior, not subject to overflows and requiring no protection.

The soil of Louisiana, scientifically considered, may be divided into the following classes:

1. Alluvial soil. 2. Marsh prairie soil. 3. Port Hudson soil. 4. Orange sand soil; and there may perhaps be added 5. Limestone soil, comprising the Mansfield and Jackson formations; but of the last I can not speak from personal observation, as I have not yet visited the northern portion of the State.

The alluvial soil may be sub-divided into Red river alluvion, confined to the Red river regions, and the Mississippi alluvion, which extends along both banks of the Mississippi, being interrupted only on the east bank by the Port Hudson deposit, which terminates below Baton Rouge. The Mississippi alluvion is also the characteristic soil of the upper Atchafalaya as far, at least, as the mouth of Bayou Courtableau.

The Port Hudson soil is composed of a fine silt, deposited upon a loamy sub-soil, and is supposed to be a fresh water formation, when the Mississippi was still an estuary of the Gulf, and Port Hudson its extreme northern limit. This soil is confined to a very small belt, comprising a portion of East Feliciana and East Baton Rouge, which is not more than ten or eleven miles wide, being approximately, though not strictly, bounded by the Mississippi and Comite rivers. It is fertile and of easy cultivation, and is especially adapted to the production of cotton and corn.

The Marsh prairie soil predominates in the Attakapas parishes, which border on the Gulf, or were formerly below the Gulf level and whose soil was formed from the sea marshes, being gradually raised above the reach of overflows. It extends within five hundred yards of the south bank of the Têche, for the soil on the north bank of the Têche is all Missisippi alluvion, subject to overflows from that river through the Atchafalaya and Grand Lake. It also comprises some of the finest lands and the most valuable sugar plantations on Bayou Barataria and the numerous other bayous and water courses in which that portion of Louisiana abounds. The agricultural capacities of those regions, have as yet not been sufficiently developed, because the shallowness of the bayous and lakes which are so numerous, that some of them have not yet received specific names, prevent easy communication by way of the Gulf with New Orleans, the centre of commerce, which can only be reached by sailing round a long distance to the Belize and up the mouth of the river. Its only direct communication with New Orleans is by Harvey's canal, which is navigable only for very light draught steamboats, and at the slow rate of traveling of three miles an hour.

The orange sand soil may also be divided into two kinds of soil. The red sandy loam soil, and the fossiliferous gravel soil. The red sandy loam soil forms a ridge along the banks of the Teche, which was never overflowed, either by the waters of the Gulf, or the waters of the Mississippi and its tributaries. This red sandy loam ridge crops out near Port Barre, where the Teche divides off from the Courtableau, and extends, in St. Landry, to the towns of Washington and Opelousas, beyond which it ceases to be visible on the surface, being covered by the prairie soil. It also forms the foundation of the Atchafalaya river bed, and the lower strata of its banks upon which the alluvion is deposited. Geologically considered this soil might perhaps be classed with a lower series of the Port Hudson formation, but with reference to its properties as soil, it is essentially an orange sand loam, mixed with limestone ingredients, and possesses sufficient fertility to render its cultivation remunerative to the planter and farmer. I am the more inclined to class it with the upper series of the orange sand period, judging from the botanical indications furnished by the prevalent summer weed, which is invariably the bitter-weed (Helenium tenuifolium) where the sandy loam, or the fossiliferous gravel soil prevails.

Near Opelousas it almost traces the line where the red sandy loam terminates and the prairie soil begins. In Rapides, on the banks of the Teche, in St. Landry near Washington, and again beyond Ville Platte and Chicotville, from whence it extends to the fossiliferous gravel soil; in St. Tammany, in Tangipahoa, in East Feliciana, every where the yellow flowers of the bitter-weed cover the surface of the soil, and in some localities, are crowding out every other vegetation; while the Port Hudson and the alluvial soils produce invariably the four-toothed Helenium (Helenium quadridentatum,) another species of the same genus, which grows there with the same rank luxuriance.

The same sandy loam, only of a deeper and almost crimsonred, crops out near Clinton in East Feliciana, where the Port Hudson soil disappears, and pine and black jack form the characteristic growth. There it covers the second subdivision of the orange sand soil, composed of a whitish or grayish clay and fossil gravel beds.

The fossiliferous gravel soil is almost devoid of those properties which render the cultivation of the Southern staples profitable. It has considerable extent in this State. The Amite river is its western limit, and it is bounded on the east by Pearl river. It embraces the parishes of Tangipahoa, St. Helena, Livingston, Washington, St. Tammany and portions of other parishes.

I do not wish it to be understood that my geological references have the least authority; I have only touched upon the geological features of the country, in order to enable me to present the subject in a more tangible form, but I did not intend to express a decided opinion on the nature of the geological formations of Louisiana, and any seeming contradictions, if there are any, between this report and that of the geologist, must be decided in favor of the geological report, which alone can be considered as authority in this matter.

Before dismissing this subject, it is perhaps proper to say that the foregoing division of the soil of Louisiana refers only to the general nature of the surface in any given region of country, without reference to the exact proportions of its chemical constituents; that the surface soil is always more or less modified by local circumstances, and varied by creeks, bayous, lakes and rivulets, which sometimes impart fertility to the lowlands, when the higher lands are nearly destitute of productive qualities, to render them valuable for tillage. Even the second series of the orange sand soils might be improved by the application of lime manure, which is ready at hand in the shell banks of Lake Pontchartrain, Grand Lake and the shell islands in the shallow waters of Barataria Bay; or they might be planted in long leaved pine for the production of turpentine, which flourishes and grows to a great height in the fossiliferous gravel sand soil, and which would render the land as valuable almost as that of a cotton and sugar plantation. Timber plantations are very common in England, and companies might be formed, possessing sufficient capital, for the purpose of appreciating these lands, by planting them in the turpentine pine, which furnishes a valuable article of commerce, of which the Carolinas have almost the monopoly.

BOTANICAL ITINERARY.

EAST AND WEST BATON ROUGE.

The botanical survey of the State of Louisiana, in connection with the geological and topographical survey, having been determined upon, I assumed its duties, in accordance with your directions, early last spring, by devoting every Saturday and the few leisure hours I could spare during the rest of the week from my other professorial duties, to botanical excursions within the limits of the parishes of East and West Baton Rouge. I visited early in April the extreme western portion of the parish of East Baton Rouge, on the Amite river. The land lying between the banks of the Mississippi and the Comite rivers, where it is not Mississippi river alluvion, is composed of the fine silt of the Port Hudson formation, and is very desirable land for the cultivation of cotton. The principal trees, which constitute the predominant growth of the forest, are the water oak, the swamp chestnut oak, the post oak, the willow oak and beech, intermixed with magnolias, hickory, locust and a few scattered tulip trees. But, passing the Comite river, the soil gradually changes; it partakes of the nature of the red sandy loam of the upper series of the orange sand formation, and being largely intermixed with lime nodules, which, in one particular place, literally pave the edge of the road with their hard and bone like concretions, the land is still good, and produces fair cotton and corn crops. On approaching nearer the Amite river, the soil becomes more sandy; the swamp chestnut oak ceases to form the characteristic growth: water and willow oaks and hickory nearly disappear, and the beech alone remains, with black jack and post oak, and an occasional magnolia. On the banks of the Amite, where the fossiliferous gravel soil is reached, the pine makes its appearance, but sparsely intermixed with oak and a few other trees. This soil is poor, and, unless heavily manured, does not yield good crops to render its cultivation profitable.

I also made several flying excursions, during the spring season, across the river, in West Baton Rouge; but while I collected many interesting specimens in the vicinity of the Mississippi river, want of time and the swampy nature of the roads prevented me from extending my botanical excursions into the interior of the parish. The soil in this parish is very rich, partaking of the nature of the Mississippi alluvion, and the sugar plantations present a far neater appearance than those on the east side of the river. The land on Grosse Tete bayou is considered the finest in the State, and is preferred to the river land, because it is less infested with the nut grass, which taxes so much the patience and industry of the planter.

RAPIDES PARISH.

After I had finished my course of instruction at the University, I made a trip up Red river, on the fifteenth of June, and visited the neighborhood of the old Seminary, which offers the richest field in the State for the collection of such plants as are generally found in sandy soils and in pine wood regions. Nature seems to be lavish in her gifts of floral adornment where she has withheld fertility from the soil. This is indeed one of the wise provisions of Divine Providence, by which the balance in the organic world is maintained, and its existence and perpetuation secured. Were all lands equally productive of such plants which contribute to the nourishment and the economic use of man, every other plant would gradually be exterminated, and would cease to hold its place in the organic world, which is composed of animals and plants, each of which is necessary to accomplish the designs for which each species has been called into existence. The disappearance of a single species, struck from the aggregate of living organizations, might extinguish, for want of proper food, many insects or birds and other inferior animals, which in their turn are indispensable links in the indissoluble chain of association, by which all the parts of the universe are bound together.

The soil of Rapides, except where the hills of the pine barrens are covered with the fossiliferous gravel or red loam of the drift period, is that of the Red river alluvion. Although different in some respects in its constituent elements from the Mississippi alluvion, it is equally productive of cotton, sugar and corn, and is as inexhaustible in its materials of fertility. It is even more highly valued by some planters, because the banks of Red river do not cave much, and the land can therefore be more effectually protected by levees.

NEW ORLEANS.

I collected in Rapides during my four days' sojourn there a considerable number of plants which bloom in early summer; but as my presence on commencement day was desirable, I returned to Baton Rouge, witnessed the exercises of the last day of the scholastic year, and a few days after the close of the session, I proceeded to New Orleans, which I selected, on account of its facilities of communication by railroads and steamboats, as the centre of my operations for this year's survey. In making an extensive botanical tour of long duration, it is indispensably necessary, in order to preserve the plants collected, to select some convenient place as headquarters, where the collector has a room, and where he is uninterrupted in his tedious arrangements by idle spectators, to enable him to determine plants which require but little research, apply the requisite pressure to dry them, and take them out from the press for several consecutive days, change the drying paper, and when sufficiently dry to place them in new sheets, where they permanently remain until they are finally disposed of and classified. This operation is far more important than the collection of plants, and requires much more time as well as patience, and more constant attention. In preparing an herbarium of botanical and scientific value, it is necessary to collect, at the same time, a great number of specimens of the same species, and to select from those the best developed, and those that show best all the parts of the plant. A considerable number of specimens of the same species must be dried, as the herbarium should contain from four to six specimens of each species, and as some of those, which are subjected to pressure, will prove worthless, it is apparent what amount of labor it requires even to dry fifty specimens, which will fill from two to three hundred drying papers. Fifty specimens might be collected in certain localities in a few days, where no collection of the same kind has ever been made, but it would require several weeks to make a final disposal of them. I make these explanatory remarks, because they refer to facts, which no one but a professional botanist could know; otherwise there might be some misapprehension with regard to my movements, as I was only able to visit seven or eight parishes during two months traveling, and only a small portion of these parishes.

On my arrival in New Orleans I immediately went to work to examine the botanical prospects of the surrounding country, and as it was entirely a new field, I found many interesting specimens, some few of which I met with no where else in my travels. I made several excursions to City Park, Bayou St. John, Carrollton, Algiers and Lake Pontchartrain.

In New Orleans and its vicinity I found the high-stemmed Rudbeckia (Rudbeckia maxima) and the Late-blooming Parthenium (Parthenium hysterophorus), and near Lake Pontchartrain grows a species of hibiscus (Kosteletzkya Virginica), with large pink flowers, which would make a beautiful ornamental garden plant.

In the immediate vicinity of New Orleans, between the river and Lake Pontchartrain, where the New Orleans and Jackson Railroad passes, the soil is that of the Mississippi alluvion, and is well cultivated in cotton, sugar and corn. North, between Lake Pontchartrain and the city, there are extensive gum, cypress and oak swamps, which are valuable only for their timber; and east and southeast are the marshes extending to Lake Borgne.

RAILROAD TRIP TO PONTCHATOULA AND AMITE CITY.

After having visited the surrounding country about New Orleans and ascertained its botanical resources, I proceeded on the New Orleans and Jackson Railroad to Amite city, which is about ten miles from the State line. Between Kenner and Frenier there is a large tract of swamp prairie, where the Rose-mallow and the Arrow-leaved Ipomaea form, as far as the eye can reach, a carpet of flowers, variegated with pink, white and crimson. A few miles from Pontchatoula, where the pine woods commence, the land on both sides of the railroad are gum and cypress, or willow and cypress swamps, which give to the country a very monotonous appearance, unless interrupted by a few miles of swamp prairies. From Bayou de Saules to Deseret's station the railroad tract is but a short distance from Lake Pontchartrain. Tangipahoa parish is, as I am told, about sixty miles long, and from three to five miles wide, and is, in point of soil, perhaps one of the poorest parishes in the State. The soil is sandy, without much intermixture of clay or gravel, and the farms are principally on the water courses, where the land is low and enriched by the deposits produced by high water, and which alone can impart some fertility to the low lands. The prevailing growth of the forest is the long and short-leaved pine, post oak and black jack, and near the Tangipahoa river water oaks and beech are not uncommon. The prevailing weed is the bitter-weed, which grows here in rank luxuriance. The gneral characteristics of the plants which flourish here are similar to those of the pine woods of Rapides. Amite city is the largest town of some note in this parish. It has a fine hotel and a number of respectable dwelling houses. I also stopped one day at Pontchatoula, and the numerous new specimens I found there well compensated me for the long walk I had to take in the heat of the mid-day's sun, for my attempt of hiring a horse had failed from the fact that the few disposable horses in town were all affected with the distemper.

BRASHEAR CITY AND FRANKLIN—ST. MARY PARISH.

As soon as circumstances permitted, I set out for St. Mary parish, for, being one of the richest of the gulf parishes, I supposed it would offer a fine field for botanizing. To travel from New Orleans to Brashear City it is necessary to cross the river in the Jackson park ferryboat, and the railroad cars start at 8 o'clock in the morning from Algiers, and reach Brashear City at 12 o'clock, M. The country through which Morgan's Texas Railroad passes is rich in alluvial soil, and the sugar, cotton and corn plantations I saw were generally in good condition; and if the crops had, in a few places, a sickly appearance, it was not the fault of the soil, but was owing to the want of steady labor to keep the fields clean from grass and weeds.

Brashear City is scattered over a considerable extent of ground, but its appearance does not correspond with its high sounding title. It is only a jumble of mean and insignificant looking one-story houses, and has not a single regularly built up street. It is, however, of some importance in a commercial point of view, for here all the cattle are landed from the Texas steamers. It is the terminus of Morgan's New Orleans and Texas Railroad, and is the point of departure of the Galveston and Rockport steamers by Berwick's Bay, which, by the by, is not an arm of the sea, as might be supposed, but a kind of inland fresh water bay interposed between the Atchafalaya river and its mouth. But notwithstanding that Brashear City is an inland seaport, and the terminus of a railroad well patronized, it is nevertheless a nondescript place, and it has not a single regularly laid out wagon road leading from there to any part of the world. I suppose that its water and railroad communication renders all other roads a superfluity. It has the Atchafalaya river and Berwick's bay on the west, and Lake Palourde lies east of it, while Bayou Bœuf empties below the town into Berwick's Bay, and thus encloses a triangular strip of land which forms an island called Tiger island.

In the outskirts of the town, along the banks of Bayou Bœuf, miles of abandoned lands can be seen, as level as a prairie, which were cultivated before the war, but are now permitted to lie idle for want of capital, and which would be productive of fine crops of sugar, cotton or corn, if properly managed. The prevailing timber trees in this region are the live oak, the water oak, the sweet gum and locust. But the trees are all thickly overhung with the funereal long moss, which, if it were not an air plant, might be supposed to prey upon the vital sap of these trees, and eat out their substance, for the greatest number of them are of a stunted and scrubby growth, and many old trees, being in a state of decay, are rapidly dying out. This sickly appearance of the timber trees is probably owing to former periodical overflows, to which these forests were exposed from the Mississippi waters. Elder bushes and Trumpet flowers, the Four-toothed Helenium and the Burdock—the two latter familiar Baton Rouge acquaintances—constitute the rank vegetation during summer, which covers the face of the land. My excursions were principally confined to within two or three miles in the vicinity of the town, on both sides of the bay. I found several specimens here which I had never seen before, nor have as yet met with in any other part of the State. The Grass-leafed Schollera grows abundantly on the edge of Berwick's Bay, and its long wiry stems, grass-like leaves and small yellow flowers cover large patches on the shallow waters of the bay, which, in the distance, appear like floating garden spots in the midst of the waters. On the banks of the Atchafalaya I found the long-flowered tobacco(Nicotiana Longiflora,) which is hardly indiginous in that part of the country, and must have been introduced from other parts of the world.

I remained in Brashear City but one day, and availing myself of the daily steamboat line up the Têche, I proceeded to Franklin, the parish site of St. Mary, a distance of thirty miles, which I reached in the evening, after six hours travel. The Têche empties into the Atchafalaya river, about twelve miles above Brashear City, and the country a short distance beyond its banks, may be considered the garden spot of Louisiana. The sugar plantations of the lower Têche are perhaps, with the exception of Cuba, the finest in the world. The country residences are for the most part neat cottage houses, commodious and well constructed, but not as elegant as the fine mansions of the coast plantations. The sugar houses are substantial and solid, but display no architectural taste. The Grand-Wood place has a fine drawbridge, or rather swinging bridge, which connects the southern with the northern bank, and from this point up to Franklin nearly every plantation has its drawbridge and landing, and the river front of the south bank seems to be entirely fenced in. The immediate banks of the Têche are composed of a sandy, red loam ridge, which I have classed with the upper series of the orange sand soil, and which is but moderately productive, and if not manured becomes exhausted in four or five years. This ridge is not more than four or five hundred yards wide on each bank of the river. The sugar lands on the south bank have been redeemed from the sea marshes, and this soil when wet is as black as coal. The humus accumulated in it for ages renders it almost inexhaustible. The land beyond the immediate north bank is composed of Mississippi alluvion, and although as fertile as the soil of the opposite side in the production of sugar and corn, the plantations are more frequently interrupted by tracts of woodland, because the backwaters of the Mississippi frequently inundate these regions through the Atchafalaya and Grand Lake; while the high and receding waters of the Gulf never reach the narrow strip of land, not more than three or four miles wide, extending from the south bank of the Teche to the impassable marshes of the Gulf. The lowest swamp lands, which are not susceptible of cultivation and are overgrown with the finest cypress this continent produces, are made available by the proprietors of plantations on the south bank for lumber, for they had the good sense of constructing fine sawmills on their plantations. The alluvial land on the north bank is also confined to a narrow strip from three to five miles wide, being bounded on the one side by the Teche and on the other by Grand Lake and other small lakes connected with it, which of itself forms an inland sea, from thirty to forty miles long.

The characteristic timber trees of the Teche country, are the live oak, the water oak, locust and hickory. The Bitter-weed (Helenium Tenuifolium) covers the red sandy loam ridge of the Teche, and appears only on the sugar lands as a stray straggler, whom accident has transplanted upon uncongenial soil. The edge of the sea marsh, near Bayou Portage, which is timberless, is rank with sedges and a new species of Hydrolea, which I have named Hydrolea Ludoviciana.

The distance from Franklin to a point on Bayou Portage, as far as it is accessible on horseback, is about three miles. It is seven miles from Bayou Portage to Red Fish Point, immediately on the Gulf of Mexico, which, on account of the impassable sea marshes, is inapproachable by land, in all the Attakapas country, and can only be reached in skiffs through the bayous which wind through the marshy lowlands, overgrown with rank vegetation, and entirely covered with water during the rainy season. I passed through strips of timber land, principally live oak, water oak and locust, and here the dwarf palmetto (Sabal Adansonii) attains a considerable height, and, if I am permitted to use the expression, forms the undergrowth of these forest wilds, with the ever present long moss hanging from the tree branches in long festoons, which gives to the scenery a semi-tropical appearance; and were it not for the musquitoes and sand flies which reign supremely in these parts, the lover of nature might enjoy for a short time the solemn stillness of the solitude, where nothing is seen but the canopy of heaven, the bright green foliage of the trees, brought into relief by the ashy gray of the long moss, the fan-like leaves of the palmetto, and the calm, still waters of the Portage. The edge of the sea marsh which forms a timberless marsh prairie, is grown up with marsh grasses and rushes, and affords valuable pasture ground for stock; and stock farms would probably pay as well as the cultivation of sugar, requiring less capital and less labor.

Franklin is a town of considerable size, containing from twelve to fifteen hundred inhabitants, is regularly built, and has many elegant private residences, which is a sure indication of the refinement and intelligence of a community. The side walks of the town are made solid by means of shells, which are brought from the lakes on the opposite side of the river.

To give some variety to my narrative, I may be permitted to state, not as a legitimate item of my report, but as a curious incident of my travels, that on Sunday night, one of the two churches being open for public worship, the whole congregation consisted of seven persons, and I was one of their number. The idea naturally forced itself upon my mind, that the people of Franklin must either be very good, having passed the praying point, or they must be retrograding in the opposite direction. Which of these alternatives is the true one, it is not important to decide.

This part of Louisiana is considered the paradise of hunters, deer abound here, and bears are very numerous in the marshes, and have been killed within a mile of town.

MANDEVILLE AND COVINGTON—ST. TAMMANY.

The Teche country, while unsurpassed in agricultural resources, did not present, botanically, as much interest as I anticipated, and as I was informed that the characteristics of soil and vegetation of New Iberia parish and higher up the Teche, were similar to those of St. Mary, I concluded to return to New Orleans, in order to branch out in another direction. I took the Pontchartrain railroad cars to Lakeport, and from there a steamboat conveyed me to Mandeville, a town in St. Tammany parish on the northern shore of the lake. Lake Pontchartrain, on a bright and sunny summer day, presents the most beautiful sheet of water in the Southern States. Near the mouth of the Tchefuncta river, the azure blue of the cloudless sky, the unstained whiteness of the distant lighthouse, and the dark green foliage of the forest trees on shore, effectually contrast the ashy gray of the rippled waters, and the scenery is altogther picturesque. Mandeville is reached in four hours' time. The soil around the town belongs to the second series of the orange sand soil. The banks of the lake are composed of pure sand, deposited upon a grayish stiff clay, colored in streaks with red and yellow hydroxide of iron. The back part of the town is located in a pine wood swamp, and during rainy weather the streets, and even the sidewalks are literally under water, and are almost impassable to foot passengers. The swamp soil is covered by a thin crust of vegetable mold, which renders the land productive for a few years of fair cotton, corn and rice crops. But the highlands are composed of pure sand, with a subsoil of clay, or the clay itself forms the surface soil. The water oak and swamp chesnut oak, post oak and sweet gum grow here to a great height, intermingled with short leafed pine, which forms, a few miles from the lake shore, almost the exclusive growth of the forest. The prevailing summer weed is the characteristic bitter weed. Mandeville is a town of respectable dimensions. It is the summer resort of New Orleans families, and is said to contain several thousand inhabitants during the bathing season. The principal street extends along the lake shore; and each residence, being generally rented for the season to the city people, has a bathing establishment, consisting of a small cabin built in the lake, at some distance from the shore, which is approached by a narrow plankwalk, protected by bannisters, constructed upon posts, and raised from ten to fifteen feet above the surface of the water. During the winter Mandeville is a deserted village, for the few proprietors and owners of houses, who remain there, live exclusively on the income derived from the rent of their houses during the summer months; and the surrounding country is too poor to support a town of a any size.

There is a colony of Indians (Choctaws) near Bayou Lacombe, about seven hundred in number, a dozen of whom—men, women and children—were emigrating from their settlement, and passing through Mandeville on their route. I was struck with the low and degrading condition of those who were once the lords and masters of this continent. Dressed in civilized rags, the once brave and lordly red man is now the mean and abject beggar of American civilization, appropriating to himself all its vices, without adopting its counteracting virtues. The squaws were carrying their papooses, or a portion of their scanty household ware, in long baskets strapped to their forehead, like cattle yoked to the plow, and thus they were wending along their weary way in true Indian file, while the men leisurely measured their steps, with their gun or perhaps a venison ham strapped to their shoulders. "O! how are the mighty fallen!"

From Mandeville I proceeded to Covington, which is about ten miles distant, the road passing through the pine woods. Here I found for the first time the Violet-flowered Stokesia *(Stokesia Cyanea)*, a beautiful composite flower, which is far more ornamental than some of the rare exotics cultivated in our gardens. I have thus far not met with it in any other part of the State. Covington is on the Bogue Felia, one of the branches of the Tchefuncta river, a navigable stream which empties into Lake Pontchartrain; but steamboats come only within three miles of the place. It is the parish site of St. Tammany. It has a few regular streets; but most of the private residences are scattered over a large area of ground, with the pine woods all around them. These are very respectable looking frame houses, generally owned and occupied by people who have small stated incomes, drawn from other sources than such the town affords. The steamboat arrives three times a week, and thus establishes regular communication between this place and New Orleans. The landing is in the midst of the woods, with no house or shelter of any kind within three miles of it. A number of carriages are punctually in attendance, at the arrival and departure of the boats, to convey passengers to and from the town. The soil is composed of sand or a stiff yellowish clay, tinted with iron; and a short distance from town the banks of the Bogue Felia are made up of layers of this yellowish clay, and are from twenty to thirty feet high. This kind of land is almost valueless, except for its timber, and for the manufacture of bricks; and there are a considerable number of sawmil.s and brickyards on the Tchefuncta, the lumber and brick being sent to New Orleans in sloops, which swarm in these waters and on the lake. At one of the sawmills, we passed, the river bank was composed of bogue iron ore in layers, which is one of the characteristics of the orange sand formation. The short leafed pine, intermixed with an occasional oak, forms the predominant growth of the woods. The botanical specimens collected here were numerous, and many of them interesting.

GRAND ISLE.

As it was impossible to reach the gulf coast in the Attakapas parishes, I availed myself of the steamboat communication which places New Orleans in direct connection with Grand Isle. I started on Saturday morning at eight o'clock, in the Col. D. S. Cage, a small steamboat drawing about eighteen inches water, from Harvey's canal, which takes its beginning a short distance from the river in the outskirts of the town of Algiers. It is six miles long, from thirty to forty feet wide, and just deep enough for a light draught steamboat to travel from three to four miles an hour. The land on both sides of the canal is wholly composed of willow, gum and cypress swamps, and is not susceptible of cultivation. There are a few settlements on the banks of the canal, consisting of a number of miserable huts surrounded by a few acres of cleared land cultivated in corn; fishing being the principal occupation of these "natives." Bayou Barataria, which is the outlet of the canal, is a beautiful little stream whose waters are of the deepest green, and it has some fine sugar plantations on its banks. On this bayou, as well as on the bayous and lakes with which it connects, there are miles of marsh prairies with not a single tree, except perhaps some stray willow to shade them, and the dusky woodland in the distance bordering the horizon. After two hours' traveling the bayou gradually widens. In some places there is a distance of a quarter of a mile which separates the opposite banks. Here the land is principally composed of marsh prairie soil, thickly overgrown with marsh grass (Spartina juncea). Barataria bayou connects with Bayou Rigolet and Little Lake, which empty in St. Denis bayou. These bayous and lakes are dotted with shell islands, some of them quite prominent for their extent and height above the level of the water. These shells might be made a valuable commercial commodity. They are the best material for making solid and substantial roads, and would serve as valuable manure to improve the pine lands of Louisiana, and they are partially used now for these purposes. But as the supply is almost inexhaustible they ought to be worked systematically and the shells transported on a grand scale. These islands are, in an agricultural point of view, as valuable as the guano islands, and to fertilize the orange sand soil shell lime is a far more useful manure than guano. This species of shell (Gnathodon caneatus) is found nowhere else in the world except in Lake Pontchartrain, Grand Lake, the waters of the Gulf, and Mobile Bay; and some enterprising capatalists of Louisiana would confer great benefit upon the State in bringing this article into market for the use of planters and farmers, and for making some of the roads of Louisiana, which are almost impassable during the winter, permanently solid, far superior to any plankroads that can be constructed.

The banks of these multifarious water courses, all formed by the receding waters of the Gulf, assume many zigzag shapes and fantastic indentations, and the water spreads and covers a vast area, forming all around numerous narrow points of land. Sometimes the marsh grass obstructs the bed of the stream, and leaves only a narrow channel, just wide enough to let the boat pass through. St. Denis Bayou has its outlet in Grand Lake, which constitutes the western border of Barataria Bay, and is about fourteen miles from Fort Livingston and seventeen miles from Grand Isle. These bayous and lakes form numerous branch bayous in every direction, which gives to the whole country the appearance of an inland archipelago, interspersed with small islands and peninsulas of every imaginable shape. The land being perfectly level, the sight is not obstructed, and the eye is struck with these ever-varied alternations of land and water, which are constantly contending for mastery; but in proportion as the bed of the Gulf is sinking, these lands will rise and become habitable to man, now occupied by the wild duck, the crane, the gull and sometimes by a stray fisherman or duck-hunter. Grand Lake connects with Caminada Bay, which is very shoal water, being in some places not more than three feet deep. It borders Grand Isle on the north. At Fort Livingston, which is situated on Bonne Terre Island, and which defends Barataria Grand Pass against marauders and smugglers, is a fort of very little importance. The work is constructed of brick, with a lighthouse, to prevent its being surprised at night in its loneliness and isolation. It has no garrison; a lieutenant and a sergeant compose the whole military force; and this is considered sufficient to hold it against any enemy, at least in time of peace. This and the adjoining islands were the headquarters of the pirate Lafitte and his robber crew.

Grand Isle can only be approached at its northern border and from Caminada Bay when the tide is high; for on its southern Gulf shore no boat can approach it on account of shoal water and three sand banks near the beach, which form breakers and in stormy weather would render navigation dangerous, even if the water were deep enough near the shore to allow a passage. At low tide the boat casts anchor within a mile and a half of Grand Isle wharf, and passengers and baggage are transported in yawls to the island.

The beach of the Gulf shore is, as might be supposed, pure sand intermixed with shells thrown out by the waves which are at all times dashing against the shore. But about half a mile from the beach the soil becomes a black muddy clay, similar to that of the sea marshes, covered with a layer of sand swept there by the overflows, which are very rare here, and are produced only by violent storms and last but a few hours. In 1854 there were several fine sugar plantations on the island, but the overflow of that year drenched, the cane with sea water, and as salt is most injurious to the cultivation of sugar, most of the cane was killed, and no attempt has since been made to revive its culture, which might be successfully accomplished by deep ploughing so as to bring the marsh soil to the surface. The sea island cotton is still produced here to a limited extent. Garden vegetables of all kinds are cultivated here with great success, and truck farms for the New Orleans market might be made profitable.

The island is nine miles long and one mile wide. The principal natural growth is the live oak and the yaupon in the form of low thickets. A few tree palmettoes are seen here and there near the beach. The live oak is low and stunted and grows only on a few ridges; the rest of the island is destitute of trees of any kind, and the inhabitants get their fuel from the drift wood floated to the Gulf shore. There is a piece of low ground near the beach, where a considerable number of barkless live oak trunks, with their leafless branches, are looming like spectres over the land and sea level, as if attentively listening, with outstretched arms, to the roaring din of the waves, incessantly advancing and receding and dashing the foaming surf against the shore.

There are a great number of small islands in the vicinity inhabited by fishermen, oystermen and duck hunters, for these waters supply the New Orleans market with the best oysters, fish and crab the sea affords, and also with wild ducks, which resort to the marsh prairies by thousands, feeding on the grain of the marsh grasses.

This secluded spot is the summer resort of many families and gentlemen of leisure from New Orleans, who avail themselves of the gulf breeze, the sea bathing and the fine table of the host, who furnishes the best the sea and land affords. But when the winds are lulled, the musquitoes, like a host of locusts, season with a drop of poison the cup of pleasure. I have collected here a few sea weeds and sea shells which are of some interest as they are the products of Louisiana waters. The specimens of the plants collected are those peculiar to the sea coast, and are found nowhere else. The greatest number of them have fleshy leaves, which is one of the wise provisions of Providence, that enables plants to grow in a dry sandy soil which holds no water, by storing away the materials of nourishment in the leaves covered with a thick and impermeable epidermis. The Goat-foot leafed Ipomœa, with thick, bright green, glossy orbicular leaves, and showy crimson flowers, is trailing all over the sand beach near the gulf shore, where no other vegetation flourishes.

Politically considered, the island forms a part of Jefferson parish, and is, it is said, ninety miles from the courthouse. In some respects this island may be looked upon as a primitive paradise of the golden age of society, for no law officer, not even a justice of the peace, resides on its hallowed ground, yet the tax gatherer and publican finds his way to this lonely sea-encircled spot, and exacts the tribute due to Cæsar.

OPELOUSAS—ST. LANDRY.

On my return to New Orleans it took me about a week to arrange my specimens, and dispose of them in such a way as to need no longer my attention; and as they occupied considerable space by their number, I left them, with my other collections, at Swarbrick & Co.'s, to be sent up to Baton Rouge at the first convenient opportunity. After having made the necessary arrangement, I started on Wednesday evening (fifteenth of August) on the regular Opelousas packet, and proceeded up the Mississippi river, down the Atchafalaya, and up the Cortableau bayou as far as the town of Washington, in the parish of St. Landry. The Atchafalaya takes its rise about eighteen miles beyond the mouth of Red river, and during low water receives its waters exclusively from that river and its own tributaries. It may therefore be considered as that branch of Red river which empties into the Gulf. When the Mississippi is high, it disgorges its overflowing tide into the Atchafalaya, which thus becomes one of its principal outlets. The back waters of the Mississippi river ascend the mouth of Red river, whose bed is sloping upward, but finding the opening of the Atchafalaya in their way with its downward sloping channel, they rush into it with great force, and swell its volume very rapidly, so that its lower banks scarcely ever escape from being inundated, unless protected by effective levees.

The land on the Atchafalaya, the west bank of which forms the boundary line of St. Landry, is composed of alluvial soil similar to that of the Mississippi river, with a clay foundation, producing fine corn and cotton, but being exposed to overflows, and being very heavy timbered, it is mostly settled by small planters, and the plantations are at some distance from the river. The Atchafalaya is broader and deeper than the Red river, and during low water its banks are in some places from forty to fifty feet high. It is navigable by steamboats throughout its whole length to the gulf. But in order to reach Washington, which might be called the inland seaport of Opelousas, it is necessary to ascend Bayou Cortableau, a navigable stream of considerable depth. Here alligators are swarming in great numbers. When they appear, swimming on the surface, the passengers amuse themselves by firing at them with pistols and rifles. But being effectually protected by their coat of mail, a few only are hit at the tender point, and they generally escape the murderous aim of their assailants by diving into deep water, where no ball can reach them. When the water is low, a sand bank, about two miles beyond the mouth of the bayou, prevents the New Orleans packets from ascending any higher, and passengers and freight are transported in flatboats over the bars to a small steamboat constructed for that purpose, which travels along slowly a distance of forty-two miles, until Washington is reached, which is at the head of navigation, and whence passengers are conveyed in hacks to Opelousas, which is six miles distant.

The land on the banks of the Cortableau is mostly low willow and cypress swamp until within a few miles of Port Barre, where the Teche takes its rise, dividing off from the bayou. There the land becomes hilly, and is composed of the same sandy red loam which constitutes the red clay ridge in which the Teche has its bed, rising above high water mark, and not extending more than a quarter of a mile on either side. This red sandy clay, which at the outskirts of Opelousas is covered by the prairie soil, again crops out beyond Ville Platte, and extends a few miles beyond Chicotville, where the pine prairies and the pine woods begin. This kind of soil, where it does not wash, is quite fertile for a few years, and the timber trees near the water courses, composed of willow, oak, sycamore, locust, post oak, red oak and hickory, are quite heavy. The bitter weed (Helenium tenuifolium) grows here in the greatest abundance. Sometimes strips of this red loamy soil are covered by black prairie soil for a short distance, and then the red clay makes again its appearance on the surface. The pine prairie land is as productive as the other prairie soil, but this being the boundary line which the waters of the Gulf only reached at very high tide, the surface soil forms only a thin crust, and wears out in a few years. The pine woods, at the edge of the prairies, are intermixed with oak, which really constitutes the principal growth, but the lands are low, and during rainy weather they are covered with water, and the country becomes for miles one continuous swamp. Beyond the edge of the prairie, about a mile and a half to the right of the Alexandria road, the land is very poor, and on both sides of the road, which runs on an elevated ridge, there are seen a series of ravines, some of them from fifty to sixty feet deep.

During the summer months the prairies are destitute of flowers, except in very low places, where the waters collect and are transformed into prairie lakes. The wire grass, which is in seed during the summer months, covers with its half-withered stems the whole surface of the prairie level. Cattle and other stock are scattered all over the open pasture grounds where the lands are not fenced in for cultivation. During the winter these prairies do not afford sufficient subsistence for a large number of cattle, as the grass dies out, and stock raisers are compelled to drive their stock during that season to the neighboring cane-brakes. But if these prairies were planted or sown in the Texas musquit grass, which is evergreen, these lands would become invaluable as stock farms. The land on Bayou Bœuf is altogether different from the prairie soil and the red sandy loam soil. It is alluvial in its composition, and contains a considerable quantity of vegetable mold, but in some localities its loamy ingredients, being nearly destitute of sand, are so stiff and unyielding that, during a long continued drought, they bake, become hard and cloddy, so that the plow and harrow can only pulverize them with the greatest difficulty.

About six miles from Chicotville there is a quarry of bluish limestone, which was formerly worked, but is now abandoned. There are also mineral springs within ten miles of it. But as circumstances prevented me from visiting these localities, I can give no account of them, or the soil and vegetation in the vicinity, from personal observation. The land between Washington and Opelousas, as well as on the Courtableau, beyond its immediate banks, is of the same character as that on the banks of the Teche. It is a cotinuation of that ridge, and produces very heavy timber, such as oak, hickory, tuliptree, sycamore, locust and catalpa, all of great size. Cotton and corn grow here very luxuriantly, and the crops had as fine an appearance as in any part of the State I had visited. On the banks of the Courtableau, at the edge of the town of Washington, there is a chalybeate spring, which pours forth a large stream of water, and deposits the iron it contains in solution, which, on exposure to the air, becomes the insoluble yellow hydroxide of iron. I was also told that on digging wells here a deposit of iron is reached within fifty feet of the surface. Here I saw, for the first time, the fig tree which bears what is called the perpetual fig. This fruit is as large as a pomegranate, and quite sweet and agreeable to the taste. It continues ripening up to frost. As the climate of Louisiana is almost everywhere favorable to the growth of this valuable fruit tree, it ought to be extensively planted. It should be set out where it is sheltered from the direct rays of the midday's sun. On the upper Atchafalaya, where it approaches Red river, the banks rise in many places to a considerable height, where the red sandy loam of the Teche becomes exposed; but when the banks are high on one side of the river, they are correspondingly low and swampy on the other, and are under water even during the dry season of summer.

Opelousas is perhaps the oldest town in the State. It was originally a military post, and has grown up from a few straggling houses to its present dimensions, which are quite respectable, considering that it has no railroad communication, and no navigable stream nearer than six miles. The private residences are old, and time has marked them with its smutty fingers. The courthouse, which ought to be a building of some note, in a parish of the intelligence and wealth of St. Landry, is a dilapidated concern, fit only to be torn down to construct a building of some taste and pretensions in its place.

Washington is a place newly built up, and exhibits much life and energy. It is of considerable commercial importance, for this is the connecting point between New Orleans and every part of the parish of St. Landry and a portion of the surrounding parishes.

PORT HUDSON AND CLINTON, EAST FELICIANA.

My botanical collections in St. Landry were extensive, not in the prairies, but in the pine woods, and that part of the parish where the red sandy loam prevails. Finding the prairies during summer an unfavorable field for botanizing, I did not extend my excursions to Calcasieu and other prairie regions, which must be visited in spring or early summer, in order to obtain specimens of their characteristic vegetation. I, therefore, determined to close up my botanical tour by stopping, on my way to Baton Rouge, at Port Hudson and Clinton, in East Feliciana.

Port Hudson, as a town, presents nothing that is attractive, except its fine view up the river, and its historical renown as one of the strongholds in the late war. The fortifications, which are nearly intact, form a prominent feature of its surroundings, and if sodded would afford delightful walks around the place—which is susceptible of considerable improvement—especially after heavy rains, when the streets are muddy, and the sticky clay retards the progress of the weary traveler who lands on these steep and hilly banks, and has to climb up an almost perpendicular declivity. As the Mississippi was very low, the composition of its banks was exposed to the view almost to its very bed. The lowest formation, upon which the Mississippi waters rest, is a compact, adhesive, blackish clay, having a tinge of blue ; above this are layers of grayish and yellowish clay, rising to a considerable height. The Port Hudson soil, composed of a fine silt, is deposited above high water on these layers of yellowish loam. It is of a brownish color, friable to the touch, and yields fine corn and cotton, equal to the Mississippi alluvion. It extends about eleven miles on the Port Hudson and Clinton Railroad, where the pine first makes its appearance. The country between the Comite and Amite rivers presents, in East Feliciana, nearly the same agricultural and botanical features as it does between the same rivers in East Baton Rouge. While on the Port Hudson soil the Four-toothed Helenium is the prevailing weed, the Bitter-weed flourishes in great abundance in both parishes, in that part which is bounded by the two rivers, and which embraces an area of about twenty-two miles in length. The long-leafed pine is but rarely seen here, but scrub and pitch pine are everywhere intermingled with oak, beech, sweet gum and magnolia. In the immediate vicinity of Clinton, the Youpon (Ilex cassine) grows by the roadside, and is found here in clumps of impenetrable thickets.

With the exception of the bottom lands near Pretty Creek and the Comite, the highlands belong to the Orange sand formation, which is manifest from the characteristic encrinitic pebbles I picked up. In many localities the soil is composed of pure sand, which here has the real orange tint from which the formation derives its name. This sand is not entirely destitute of lime, for some of the fossil pebbles are not perfectly silicified, and still retain a portion of their carbonate of lime, which is friable and easily intermixes with the soil. The subsoil is composed of clay of the deepest ocherous red, which in some of the railroad cuts make up layers from ten to fifteen feet high, and which are so compact, that during the war, many a soldier cut his name in the clay to perpetuate his memory, if not in columns of brass or marble, at least in banks of solid and endurable clay. These pine lands are cultivated to a considerable extent, and I was assured, that besides producing fair corn crops, they yield half a bale of cotton to the acre.

Clinton is a town of some size, regularly laid out and well built up. It has about fifteen hundred inhabitants. It contains a considerable number of neat private residences, besides the courthouse and the Masonic building, which are edifices of some pretensions.

The people of Clinton have, evidently, much public spirit, and they promise themselves a great deal, by way of improvement, whenever the long expected railroad extension from Baton Rouge is effected.

The flora of East Feliciana presents considerable interest. Many specimens have been found there, which thus far, have not been met with in any other part of the State. Here grows the Anise tree, (Illicium Floridanum,) with its beautiful crimson flowers and evergreen leaves. This small tree would form one of the finest ornamental shrubs in our gardens. On the Comité river, grows in a wild state, in the midst of the woods, the Jerusalem cherry, (Solanum pseudo capsicum,) a shrubby plant, much cultivated. It is not probable that it is indigenous, but must have escaped from the gardens, which are, however, a considerable distance from the locality where it grows wild.

CONCLUSION OF ITINERARY.

After I had completed the botanical survey of East Feliciana, it being only one week to the commencement of the session, I deemed it most prudent to return to Baton Rouge, to enable me to take proper care of the collections I had made, to arrange them, determine and classify them; a labor which requires much time, considerable patience and much research. On comparing the results of my various excursions, I found that my collections, without exhausting the materials of summer vegetation, were varied, rich and highly interesting. The sea coast plants, which could not be obtained at any other locality, are, by themselves, worth all the trouble I had taken, and the expense I had incurred.

ECONOMICAL, ARTISTIC AND MEDICINAL USE OF PLANTS COLLECTED.

A report which is intended to be disseminated among the people of the State, should not only be a scientific contribution, pointing out all the striking points of interest, which relate to the particular science of which it treats, but it should contain information of practical value, which might be useful, not only to the farmer and mechanic, but to every person who does not live exclusively for himself, and endeavors to exert his faculties for the advancement of civilization, and the promotion of the happiness and well being of mankind.

I therefore consider it a matter of paramount importance to point out the practical uses of the plants already collected, and thereby show by irrefutable facts the practical bearing which botanical investigations have in all the pursuits of life, in the arts, mechanics, agriculture, medicine, and even the domestic affairs of the household. It also shows that nearly all the accumulated wealth of every country is derived from the vegetable kingdom, and that it is a subject well worthy of our study, and deserves our serious consideration.

TREES, SHRUBS AND VINES.

The Large-flowered Magnolia (Magnolia grandiflora) is widely diffused in Louisiana. Its glossy evergreen leaves and its large odorous flowers render it unequalled as an ornamental tree. The fine magnolia groves of some plantations present great natural beauty, and these rural spots should never be desecrated by the axe of the woodman. Every planter has a sufficient quantity of land for cultivation without encroaching upon the clumps of magnolias, which ought to be preserved as pleasure grounds or parks. This tree grows in cool and shady places, where the soil is covered with mold, and in pine barren swamps enriched by decayed vegetation. Its bark was used by the Southern Indian in cases of intermittents. Its wood, remarkable for its whiteness, is too soft to be employed in architecture or in cabinet work.

The Tulip tree or the White poplar (Liriodendron tulipifera) has been met with in East Baton Rouge and St. Landry, and other parishes. It delights in deep loamy fertile soils, in rich bottoms along the rivers or borders of swamps. Its wood, though light, is sufficiently compact to be used in cabinet work. It may be employed as a substitute for pine and cedar in the construction of the interior work of houses. When boards made of this tree are perfectly dry, they take paint well, and admit of a brilliant polish, and on this account it is stained in imitation of mahogany. It is also useful for bridges, as it unites lightness with strength and durability. The bark of this tree is strongly tonic and antiseptic. The aromatic principle seems to reside in the resinous part of the substance of the bark, and acts as an internal stimulant. The Indians employed it in the cure of intermittents. But its highest value is its beauty in a living state. Its angled and lobed leaves, and its large tulip-like orange flowers, its spreading and wide branched proportions, render it one of the finest forest trees that grow on this continent.

The Golden-fruited Orange tree (Citrus aurantium) grows on the coast below New Orleans. It is believed to have been originally a native of the warmer parts of Asia, but has long since been acclimated in the southern part of Louisiana. It is cultivated for profit, and the orange plantations of the lower coast are the most valuable cultivated lands in the State. I have no data to estimate the annual value of the orange crop of Louisiana. But if the sources of information were known, a compilation of statistics with regard to this important branch of agriculture in our State would be extremely interesting. The wood of the orange tree is hard, compact and flexible, slightly odoriferous, and susceptible of being polished. It is used to make dressing cases and other articles of fancy work, and the straight young shoots are manufactured into walking canes.

The Small Buckeye (Aesculus Pavia) is a low shrub about six or eight feet high. It is found almost everywhere in Louisiana in fertile soil. In this State the shrub is too small in size, and no particular use is made of the wood. We are told by Elliot, and such is the popular belief, that the bruised branches and powdered seeds have the property of stupefying fish. When the water of small ponds is impregnated with them, the fish rise to the surface almost lifeless and may readily be taken with the hand. He also tells us that the root is used as a substitute for soap in washing woolen clothes. Its fine clusters of red flowers, which appear in early spring, and the graceful arrangement of its symmetric leaves, recommend it as an ornamental shrub of the gardens.

The Pride of China or China tree (Melia Azœderach) is said to be a native of Persia, but is now naturalized in our climate. It grows in great luxuriance, and its dark green and profuse foliage renders it very valuable as a shade tree. It is, however, objected to by some persons, on account of its berries which it throws off continually, from the time they attain maturity to the period it begins to bloom again in the spring. Its wood is considered strong and durable, and has been employed in the manufacture of pulleys. The fleshy part of the berry yields a fixed oil, which has anthelmintic, narcotic and stimulant properties. The leaves are universally used in India for poultices, and both the flowers and seeds are stimulants. The berries have been pronounced as poisonous by Arabian physicians, but in this country they are eaten by children without injurious effects. The bark of the root, when geeen, has a bitter nauseous taste, yielding its active principle to boiling water, and may be employed as an emetic, and is considered an efficient vermifuge. In Persia an ointment is made for the cure of some cutaneous eruptions by mixing the leaves with lard. The nuts are often bored by monks and strung into beads. Hence its name "bead tree." "Pater nostri di San Domenico."

The Prickly Ash (Zanthoxylum Carolinianum) grows from twelve to fifteen feet high, and is found on the banks of small water courses. It branches out with a regular bushy head, at some distance from the ground; and when in bloom is crowned with a cluster of yellowish green flowers. The tree, when young, is armed with powerful prickles, which are angular and sharp at the point. The bark and capsule are of a hot and acrid taste, and when taken internally, act as a powerful stimulant. They are sometimes used to relieve the toothache, hence its name " toothache tree." They are also employed for curing intermittent fevers and rheumatism. The American Indians were acquainted with the medicinal properties of this tree. They extracted from the berries a salivating substance, and used the decoction of the plant to produce perspiration.

The Three-leaved Ptelea (Ptelea trifolia) is found in most shady places and on the borders of woods. I have met with it in this State in East Baton Rouge. It is of no particular use, but is worth cultivating as an ornamental shrub, both on account of its leaves, which are arranged in whorls of threes; and on account of the beauty of its fruit, which appears in clusters of greenish yellow flat winged seeds.

The American Holly (Ilex opaca) is a beautiful evergreen tree, sometimes growing to the height of from thirty to forty feet. It is widely diffused all over Louisiana, in shady places and on the edge of swamps, where the soil is cool and fertile. The wood is compact, heavy, of a fine texture, and susceptible of brilliant polish. It is principally used for inlaying mahogany furniture, and it is subjected to the turning lathe to make of it small druggists' boxes and small screws. When perfectly seasoned it is extremely hard and inflexible, on account of which it is well adapted for pullies of ships. The bark may be employed in making bird lime. Medicinally it is an emetic. The berries, if taken in sufficient quantities, excite vomiting.

The Yaupon (Ilex Cassine) is generally a low tangled shrub. It flourishes best in sandy soil. It has small evergreen leaves, clusters of greenish-white flowers in the spring, and bears red berries, resembling currants, which remain on the tree until new flowers make their appearance. It grows up, if properly trimmed, to a tree of small size, which is both elegant and ornamental, on account of its red fruit, intermixed with its glossy bright green leaves. The tangled, low and impenetrable thickets it forms in some localities, suggested the idea to me, that it would make a live hedge far superior to any now in use. It surpasses the Cherokee rose, which grows too high and covers too much ground with its spreading and rooting branches. It excels the Osage orange, because it branches from the base of the stem, and its growth being naturally stunted, it requires but little trimming. Besides, it is far more impenetrable and produces less shade than the leafy top of the Osage orange, and it flourishes in poor soil, of which sand makes the principal ingredient. If some enterprising planter would make some experiments with the yaupon as a hedge plant, he might confer a great benefit upon the planting communities of the prairie regions where wood for fencing purposes is not easily obtained. Its leaves are used as a tea, being almost equal, if properly prepared by roasting, to the Paraguay tea, which is derived from the leaves of a species of the holly, Ilex Paraguariensis.

The Poison Vine (Rhus radicans) rises to a great height by adhering to trees with its strong rooting fibres, which it throws out from its stem. The leaves are ternate entire, or lobed and toothed. It has greenish yellow flowers, and bears a fruit of greenish white berries. The juice, when applied to the skin, frequently produces inflamation and vesication, and it is the popular belief that a volatile principle escapes from the plant which produces, in certain persons, when coming near it, a troublesome erysipeloid affection, particularly of the face. The leaves, among other substances, yield tannic acid, and a volatile alkaloid, on which, it is pretended, its poisonous properties depend. The leaves are stimulant and narcotic; they act as an acrid poison, and produce a stupefying effect upon the nervous system.

The Dwarf Sumac (Rhus copallina) is a low shrub from five to eight feet high, and grows very abundantly in Louisiana. It bears greenish flowers, and its fruit appears in clusters of red berries which are slightly acid to the taste; the leaves contain an abundance of tannic acid, and they are sometimes collected for tanning purposes.

The Flowering Locust (Robinia pseudo-acacia) occupies the first rank as an ornamental tree, on account of the beauty of its foliage, and its clusters of white flowers. It is but rarely met with in Louisiana in a wild state. Its leaves contain much nourishing matter, and have been used as a substitute for clover, as food for cattle; but it must be cultivated for this purpose. The roots are very sweet and afford an extract similar to liquorice. The flowers, when medicinally employed, have anti-spasmodic properties, and when distilled furnish an agreeable and refreshing syrup, which, if drank with water, quenches thirst. The timber of the flowering locust is esteemed by shipwrights for the upper and lower parts of the frame of vessels. It is considered as durable as live oak and red cedar, being lighter than the former, and stronger than the latter. On account of the hardness of the wood when seasoned, and its luster when polished, it is extensively employed in cabinet work. We are told that the American Indians make a declaration of love by presenting a branch of this tree in blossom to the object of their attachment.

The Canadian Judas Tree (Cercis Canadensis) is a handsome shrub or small tree. The flowers, which appear before the leaves, are of a light purple, and are acid to the taste. The wood is very hard, elegantly veined, or rather waved, with black, green and yellow spots. When seasoned it is susceptible of a beautiful polish. The bark and young branches are used to die wool of a nankin color.

The French Canadians use the flowers in salads and pickles. The flower buds and pods would undoubtedly be excellent as pickled preserves.

The native country of the Peach tree (Amygdalis Persica) is not known. It was introduced into this country by the first European settlers at the close of the sixteenth century. I have seen peach trees grow spontaneously in Louisiana in the midst of the woods, which had probably sprung up from stray kernels dropped by some huntsman. The peach tree is principally valued for its delicious fruit, and as Louisiana has a similar climate as that part of Asia, where it flourishes best, the peach attains its highest perfection in this State, both for size and flavor. Its wood is compact and of a roseate hue, and is susceptible of fine polish. It is little used in the arts. A color may be extracted from it called rose-pink. Its leaves yield, by distillation, a volatile oil of a yellow color, containing hydrocyanic acid. Its bark, blossoms and kernels also contain the same poisonous principle.

The Common Plum tree (*Prunus Communis*) was introduced from Europe at the earliest period of the colonial government. Its fruit is pleasantly acid to the taste, and is sought after as one of the early fruits of the season. The wood is hard, close, compact, beautifully veined, and is susceptible of fine polish. The texture is silky, and when washed with lime water its glossiness is heightened, which may be preserved by the application of varnish. It is much in demand for the manufacture of musical instruments.

The Wild Cherry tree (*Prunus Virginiana*) grows here to the size of a small tree. It is a fine ornamental tree when its spikes of white flowers are fully expanded. Its wood is of a dull light red tint, which deepens with age. It is compact, fine grained, and takes a brilliant polish. When chosen near the ramification of the trunk it rivals mahogany in beauty. It is often employed for making felloes of wheels. The taste of the bark, especially that of the roots, is aromatic and bitter. It is a useful tonic, and possesses in some degree narcotic and antispasmodic properties. Dr. Barton informs us that the leaves are poisonous to cattle. The fruit is employed to make a cordial by infusion in brandy, with the addition of sugar.

The Wild Orange tree (Prunus Carolinianes) is a beautiful ornamental tree, and I have only met wi h it in this State in a cultivated state, but it grows wild in North Carolina. Its wood is fine grained, and of a roseate hue, but its scarcity has prevented its employment in the mechanic arts. Michaux tells us that a spirituous liquor may be obtained from the bark, and we are informed by Elliot that its leaves are very poisonous, destroying cattle that feed freely upon them.

The Pomegranate tree (Punica Granatum) is indigenous in Persia, Japan, and various parts of Asia, but has long since been naturalized in Louisiana. A syrup is made of its pulp, as well as the dried flowers, which is used as an astringent. The rind of the fruit has been employed as a substitute for galls in the manufacture of black ink. The natives of India make use of the bark of the root for the expulsion of the tape-worm, a property well known to Dioscorides, The fruit is pleasantly acid and quite agreeable to the taste. Its flowers are of a bright scarlet and of large size, and render the tree, when in bloom, quite ornamental.

The Flowery Dogwood (Cornus Florida) grows for the most part on the borders of swamps, and in rich soil, and is found in abundance in East Baton Rouge. It is the white four-leaved involucre, which contains a cluster of greenish blossoms, that constitutes the chief beauty of the tree when in flower. The wood is hard, compact, heavy and fine grained, is susceptible of brilliant polish, and may be substituted for numerous purposes to which box-wood is applied. It is used sometimes by farmers for harrow teeth and for hames of horse collars, but being liable to split, it should never be wrought till it is perfectly seasoned. The cogs of wheels are made of the young shoots, and the forked branches are converted into yokes, which are put on the necks of hogs to prevent them from breaking into inclosed fields. The bark may be substituted for galls in the manufacture of ink. From the bark of the more fibrous roots the Indians obtained a scarlet dye. An infusion of the flowers was used by them for the cure of intermittents. The bark of the stem, as well as the root, is employed as a tonic and astringent. It has occasionally been substituted for Peruvian bark in intermittent fevers, and has frequently been successful.

The Sorrel tree (Oxydendrum Arboreum) grows only to a small size, where it has been met with in this State. Its numerous spikes of urnlike white flowers, at the beginning of summer, render it somewhat an object of attraction. The wood is soft, of a pale rose color, and is unfit for use in the arts and for fuel. The leaves have a pleasant, acid taste, and are frequently made use of by hunters to allay thirst; and they form in decoction a grateful and refrigerant drink in fevers.

The Bignonia like Catalpa (Catalpa Bigonoides) is found growing wild in the parish of St. Landry and other parishes. It is considered a fine shade tree, on account of its large leaves, and is planted for that purpose. Its flower clusters, which appear in the spring, are large and very showy. Its wood is remarkably light, of a very fine texture, and takes a brilliant polish. Its color is of a grayish white, and when properly seasoned is very durable. It is sometimes used for posts and rural fences, and is employed in cabinet making. If a portion of the bark of the catalpa be removed in the spring, a venomous and offensive odor is exhaled. The bark is considered as possessing tonic and antisepetic properties, and has been used as an antidote for snakebite. The flower and seed are extolled as being a sovereign remedy against asthma.

The Sassafras tree (Sassafras Officinale) is indigenous to, and grows almost everywhere in the United States. The wood of the young tree is white and tender; but in trees which exceed fifteen to twenty inches in diameter it is of a reddish cast, and of a more compact grain. It is, however, of little value as a timber tree, where strength is the object. But if the wood is stripped of its bark, it resists for a considerable period the progress of decay, and on this account is employed for posts and rails of rural fences. It is also sometimes used for joists and rafters in the construction of houses, and it is said to be secure from the attacks of insects, an advantage attributed to its odor. The wood imparts to wool a very durable orange color. Medicinally, the wood, bark and roots of the sassafras are held in esteem as a stimulant and sodorific. It is used to improve the flavor of more efficient medicines, and to render them more cordial to the stomach. Sassafras pith abounds in mucilaginous matter, which readily dissolves in water. This mucilage is much employed as a soothing application in inflammation of the eyes, and forms a useful and agreeable drink in catarrhal and other diseases. The bark of the root yields a great quantity of essential oil. An agreeable beverage is formed with the aid of young shoots, and the root bark, known by the name of "root beer," which forms a salutary and cooling drink during the summer months.

The Red-fruited Mulberry (Morus Rubra) grows in East Baton Rouge and many other parts of the State. The perfect wood is fine grained, compact, though light. It is of a yellowish hue, approaching to lemon color. It possesses strength and solidity, and when properly seasoned is almost as durable as that of the flowering locust. It is employed in dockyards, in the construction of both the upper and lower frames of vessels, for knees and floor timbers. It is also used for posts and rural fences. The fruit is dark red, and has an agreeable flavor. It forms a refreshing and grateful drink, well adapted to febrile diseases. A syrup is made of their juice and used as a pleasant addition to gargles in inflammation of the throat. We are told by Du Pratz, in his history of Louisiana, that many of the Indian women wore cloaks made of the lint of the mulberry tree. They stripped the bark from the young mulberry shoots which rise from the roots. After having been dried in the sun, they beat it to make all the woody parts fall off, and then gave to the threads that remained a second beating, after which they bleached them by exposing them to the dew. When they were whitened they spun them to the coarseness of pack thread, and then wove them by stretching a cord on two stakes, fixed in the ground, and fastening double threads of bark to this cord to form the warp, they interwove the filling, and thus made themselves a species of cloak upon this very primitive loom.

The Black-fruited Mulberry (Morus Nigra) is supposed to be a native of Persia; but it is naturalized in the United States, as a valuable shade tree. Its flowers are diœcious, male and female flowers being on separate trees, very few trees bear any fruit. The wood is of little use except as fuel. The roots are considered as an active vermifuge.

The Osage Orange (Maclura Aurantiaca) is indigenous in Arkansas, Texas and Missouri. It is employed for hedges and live fences. The wood is of a bright yellow color, and is said to afford a yellow dye. It is solid, heavy and durable, uncommonly fine grained and elastic, and on account of this last property, it has been used for bows by all the tribes of Indians of the regions where it abounds. Hence its name of "Bow wood." It receives a beautiful polish of the brilliancy and appearance of satin wood. The bark yields a fine white fibre, which might be converted into thread and a beautiful woven tissue.

The Common Fig tree (Ficus Carica) is indigenous to Western Asia and the shores of the Mediterranean. In Louisiana the fig tree grows most luxuriantly, and produces fruit of the finest quality. The sapwood is extremely light and tender and of a white color, and is used for making whetting instruments, from its facility of receiving and retaining the emery and the oil, that are employed in sharpening smiths' tools. The heartwood, which is yellow, loses a great deal of its weight in drying, but by that process it acquires so much strength and elasticity, that the screws of wine presses are made of it. The charcoal has the valuable property of consuming very slowly. The leaves and bark abound in a milky acrid juice, which has been applied to the skin to raise blisters and destroy warts. Medicinally the fruit is considered nutritive and demulcent, and when roasted or boiled, it is sometimes used as a cataplasm applied to gum-boils. The fig tree is said to have the singular faculty of rendering raw meat tender, when hung beneath its shade.

The Cork-winged Elm (Ulmus Alata) is frequently met with in East Baton Rouge and elsewhere in this State. The wood is fine grained, more compact and heavy than that of the American elm. The heart wood is of a dull chocolate color, and always bears a great proportion to the sapwood. It is used for the naves of coach wheels, but it is not particularly appropriated to any other use.

The White Oak (Quercus Alba) attains under favorable circumstances a magnificent size. It is highly valued for its timber; its wood being extremely tough, durable, and elastic, is extensively employed in ship building. It is also split into thin strips for making cotton baskets and the bottoms of chairs. The bark has a rough, bitterish, astringent taste. Its medicinal properties depend on the tannin it contains.

The Black Oak (Quercus Tinctoria) is one of the finest trees of the oak family. Its bark is more bitter than that of any other species of this class, but it is less frequently used for tanning purposes on account of the red color it imparts to the leather. It contains a coloring principle called quercitrine, which is capable of being extracted by boiling water, and is used to die wool and silk of a brownish yellow color. Medicinally considered oak bark is astringent and tonic. Its decoction is used as a bath when the stomach is so much disordered as to refuse to receive medicines. It is also employed as a poultice in gangrene and mortification.

The Live Oak (Quercus Virens) is a beautiful tree which grows most abundantly in the lowlands of the southern and gulf regions of Louisiana. On account of the dark green color of its evergreen leaves it is perhaps the finest shade tree the vegetable kingdom can boast of. This oak freed from the outer wood and thoroughly seasoned will endure an unknown period of time in buildings or mach nes. Its wood, being most elastic and durable, is superior to every other forest tree, and is particularly sought after by shipbuilders. Its timber forms a valuable commercial commodity, equaled only by the teak of India.

The Long-leafed Pine (Pinus Australis) is a lofty and majestic tree indigenous in this State. It grows in dry, sandy soils, and is found in the pine lands in Rapides, Tangipahoa, and other parishes. Its timber is valuable not only as fuel, but as lumber in domestic architecture and ship-building. It yields an abundance of turpentine, and the Carolina pine supplies a sufficient quantity of that article, not only for home consumption, but also for exportati n.

The Cypress (Taxodium distichum) is one of the most valuable timber trees indigenous in the Southern States. It grows in great abundance everywhere in the swamps of Louisiana. Its beautiful foliage and its lofty and elegant form would recommend it as one of the finest ornamental trees if it grew in any other but swampy soil. Its wood is extremely porous and light, and when properly seasoned is most durable. On this account it is employed in naval as well as civil architecture.

The Sweet gum (Liquid amber styraciflua) is very abundant everywhere in the lowlands of Louisiana. When wounded, a balsamic juice flows from its trunk, which is of the consistence of honey, is of a yellowish white color, and has a balsamic odor. It has been erroneously called liquid storax, which it resembles in its properties. It is sometimes chewed by children, in order to sweeten their breath. The bark is astringent, and has been employed in the form of a syrup. The timber is valuable as fuel, and is also used for lumber where pine and cypress are scarce.

The Black Walnut (Juglans nigra) is met with in Louisiana in rich soil. Its wood, though neither strong nor compact, is extensively used in cabinet work, on account of its durability and the high polish it takes, and its exemption from the attacks of insects. The kernel of the fruit furnishes a grateful article of food. The bark is used for dying wool a dark brown color. The decoction of the bark and leaves has considerable medicinal properties.

The Hickory (Carya glabra) is one of the finest forest trees of the Southern States, and grows only in rich soil. The wood of the hickory is well known for its compactness, its toughness and durability. It is much employed for posts, and as heat-producing fuel it can not be surpassed. The infusion and tincture of the bark have been used as astringents, and have been administered in intermittents with success. Chewing the inner bark has been said to be a sovereign remedy for dyspepsia. The nut furnishes an agreeable article of food.

The Black Willow (Salix nigra) grows everywhere in the State on the banks of streams. Its wood, when exposed to constant atmospheric changes, speedily decays, but when thoroughly seasoned and kept perfectly dry, will last for centuries. It has not been used in the arts, except for making charcoal. The young shoots were employed by the Indians for the manufacture of baskets and other wickerwork. The root has tonic properties, and is used by country people for the prevention and cure of intermittents.

The Rose-bay tree (Nerium oleander) with its bunches of rose flowers is one of the finest ornamental shrubs that grows in the gardens of Louisiana. The powdered bark is said to be poisonous, and the peasants of the south of France, where it grows wild, employ it as a poison for rats, and death is said to have occurred from eating food roasted by the oleander wood. The leaves, boiled in lard or oil, yield an ointment which is considered efficacious against insects.

The American Beech (Fagus ferruginea) is a handsome tree, and grows most abundantly in low wet soil, near the rivers and bayous. Its nuts afford nutritious mast for hogs, but it has little value as a timber tree, on account of the hardness and brittleness of the wood.

The Fragrant Olive (Olea fragrans) is cultivated in the gardens on account of the delicious odor of its yellowish flowers. The odor of tea leaves, cultivated in China, is improved by mixing with them the flowers of this tree, which are afterwards separated by sifting or otherwise.

The Misletoe (Phoradendron flavescens) is a parasitic shrub which grows on the oak, the elm, the sweet gum and other trees. The berries are white and are said to be poisonous. They contain a glutinous sticky material, which surrounds the seed, and which in Europe is used in the preparation of bird-lime. The plant was considered sacred by the Druids of Britain. They looked upon the oak as the residence of the Almighty. The fruit of the misletoe, a parasite of the oak, was thought to contain divine virtue, and to be the peculiar gift of heaven. It was sought for on the sixth day of the moon with the greatest earnestness and anxiety, and when found it was hailed with rapture and joy. As soon as the discovery was made, the arch druid, attended by a crowd of people, ascended the tree, dressed in a white robe, and, with a consecrated knife, cropped the misletoe from its fixed support. Having secured the sacred plant, he descended the tree; bulls were sacrificed, and the deity was invoked to bless his own gift and render it efficacious in those diseases in which it should be administered. They esteemed it a kind of panacea, a universal remedy in all diseases.

The Common Lilac (Syringa vulgaris), is a shrub cultivated in the gardens. Its leaves and fruit have a bitter and acid taste, and have been used as a tonic and febrifuge. In some parts of France they are employed by country people in intermittent fevers.

The Canada Elder (Sambucus Canadensis) is a well known shrub' and grows every where in the United States. The flowers are gently excitant and sudorific, but are used only in the form of poultices, fomentation and ointment. The berries have diaphoretic properties, and have been employed in rheumatic and eruptive affections. The inner bark is used in dropsical complaints.

The Virginia Creeper (Ampelosis quinquefolia) is a running vine which attaches itself by its rootlets to trees and walls. In autumn its leaves turn bright crimson, and they have been used as an alterative tonic and expectorant. The bark and twigs have been recently recommended as a remedy in dropsy.

The Arbor Vitæ (Thuya occidentalis) is a beautiful ornamental tree, indigenous in North Carolina, and is cultivated in all the gardens in the United States. The leaves or small twigs have an agreeable balsamic odor, especially when rubbed, and a strong camphorous bitter taste. They have been used in the form of decoction in intermittent fever, and also in scurvy and rheumatism. A volatile oil, obtained by distillation from the leaves, has been successfully employed as a vermifuge.

The Partridge Berry (Mitchella repens) is a small evergreen, trailing shrub, with whitish, fragrant flowers, and a scarlet edible fruit. The whole plant has medical properties, and is said to be employed by Indian squaws to facilitate parturition. It is tonic and astringent.

The Yellow Jessamine (Gelsemium sempervirens) is one of the most beautiful climbing shrubs of the Southern States. It ascends lofty trees, and forms leafy bowers, extending from one tree to the other; and, in its flowering season, in February and March, it perfumes the atmosphere with its delicious odor. The leaves are perennial, and the flowers are large, tubular, and of a bright yellow color, and are said to be poisonous. The root is medicinally used. Its medical virtue has been accidentally discovered by a Mississippi planter, who, being affected by a febrile disease, ordered his servant to dig up a certain kind of root in his garden, and to prepare a tea from it, The servant dug up, by mistake, the root of the yellow jessamine. boiled it into a tea, and administered it to the patient, who was soon afterwards affected with nausea and muscular debility, but these effects gradually subsided, and with them the fever. Since that time the root has been employed in intermittent, remittent, typhoid and yellow fever, in inflammation of the lungs, and other diseases.

HERBACEOUS PLANTS.

MEDICINAL, ORNAMENTAL AND ECONOMICAL.

The Butterfly weed (Asclepias tuberosa) does not, like the other species of asclepias, emit a milky juice when wounded. Its root, which is perennial, is irregularly tuberous, branching, and has an acrid, nauseous taste. Medicinally it is diaphoretic and expectorant. It has been administered in pleurisy and pneumonia, and may be taken in powder, infusion, and decoction.

The Long moss, (Tillandsia usneoides), which is not a moss as its name imports, but bears a regular, small, greenish flower, and belongs to the order of flower-bearing plants. It has been considered for ages as useless, giving to the forest where it abounds a funereal aspect. It has recently become an important article of commerce, as a substitute for horsehair in the manufacture of matresses. After the outer covering of the flexible stem has been rotted off by exposure in the open air, there remains a black hairlike bundle of fibres, which has all the strength and elasticity of horsehair without its animal odor. It is afterwards hackled like flax, pressed in bales, and sent to New York, the great emporium of the commercial commodities of this continent. It is exclusively a natural production of the Southern States, and its growth ought to be fostered and its collection regulated by law. It is an air plant, draws its nourishment from the atmosphere, and neither exhausts the soil nor injures the trees to which it attaches itself.

The Indian fig prickly-pear (Opuntia ficus Indica) has been found on the Têche near Franklin, St. Mary parish, but it grows in great abundance in the sandy soil north of Lake Pontchartrain. Near Mandeville and Covington it covers a considerable extent of ground, and reaches the height of from four to five feet. It bears a large yellow flower, and produces a pear-shaped reddish fruit, which contains a slightly acid but extremely pleasant juice, having the appearance of red wine, and stains the fingers with its carmine dye. Were it not for its numerous impalpable stings with which it is armed, and which render its fruit somewhat forbidden fruit, its scarlet red juice might probably be converted into a delicious beverage, equal perhaps to some of those artificial wines manufactured in Cincinnati and elsewhere, and sold in the South as genuine champagne. Nor is it impossible that its coloring principle might be of some value, if means can be found by which it can be fixed.

The Worm-seed or Jerusalem oak (Chenopodium anthelminticum) grows every where in waste places in Louisiana. It has been introduced from tropical America. It has a strong peculiar, offensive and yet somewhat aromatic odor, which it retains when dried. All parts have medicinal properties, but the fruit is considered the most efficacious. It is a favorite vermifuge, and is most conveniently administered in powder mixed with syrup.

The Golden-flowered Star-grass (Aletris aurea) grows in the pine barrens. The bitter principle of the root of this plant is extracted by alcohol; and having strong tonic properties, it is advantageously taken as bitters in febrile debility.

The Balsam apple (Momordica balsamina) an herbaceous vine cultivated in the gardens, is a native of the East Indies. Its fruit, which is of graceful form and of a beautiful orange yellow tint, was formerly highly esteemed as an application to wounds, and is still in use for that purpose among country people. Infusing the fruit, deprived of its seed, in olive oil, it forms a liniment which is applied to chapped hands, burns, old sores, and the mashed fruit is employed as a poultice.

The fruit of the Pumpkin vine (Cucurbita pepo) is well known to farmers as excellent food for cattle. Its seed has of late acquired considerable reputation for the expulsion of the tapeworm, for which it is administered in the form of a paste in the quantity of an ounce and a half of the seed, mixed with an equal weight of sugar.

The Flowering Spurge (Euphorbia corollata) is very common in the sandy pine lands of Louisiana. The root alone possesses some medicinal value. Its active principle is taken up by water and alcohol, and remains in the extract obtained by the evaporation of the decoction or tincture. In a full dose it is a certain emetic. In smaller doses it is diaphoretic and expectorant.

The American Centaury (Sabbatia angularis) grows most abundantly in the pine woods and in sandy soil. Its numerous rose colored flowers, which expand late in June, render it a beautiful ornamental plant, and it deserves a place in the gardens. All parts of the plant have a strong bitter taste. Alcohol and water extract its bitterness. It has tonic properties similar to gentian, and has been popularly employed as a preventive remedy in our autumnal remittent and intermittent fevers.

The Sweet flag or Florentine orris *(Iris Florentina)* which is cultivated and constitutes quite an ornament in our gardens, is a native of Italy. The root, which is known in commerce by the name of orris root, has not only medicinal properties as an emetic, but is valued on account of its agreeable odor. It is occasionally used to conceal an offensive breath, and it enters into the composition of tooth powders.

The Common White Lily *(Lilium candidum)* is a beautiful ornamental garden flower, and is indigenous in Syria and Asia Minor. The bulb, which consists of imbricated fleshy scales, has a peculiar, disagreeable, somewhat mucilaginous taste. In a recent state it is said to have been successfully employed in dropsy. Boiled with water or milk. it forms a good emollient cataplasm, much used in popular practice. The flowers impart their odor to oil or lard, and an ointment or liniment prepared from them is used as a soothing application in external inflammation.

The Wild Onion *(Allium Canadense)* has been found nowhere else except in West Baton Rouge. The medicinal effect of the bulb of the wild onion is stimulant. It may be used in catarrhal affections of children, and in nervous and spasmodic coughs, in the form of a syrup. When bruised and applied to the feet, it is useful in febrile complaints of children, by quieting restlessness and producing sleep.

The Cardinal Flower *(Lobelia cardinalis)* grows everywhere in Louisiana, in low marshy lands. Its showy crimson flowers render it well worthy of cultivation as a garden plant, far more attractive than some of the worthless exotics that fill the greenhouses. Its root is supposed to possess anthelmintic properties.

The Sheep Sorrel *(Oxalis stricta)*, with its delicate yellow flowers, is very widely diffused in the United States. It has an agreeable, sour taste, which is due to oxalic acid, combined with potash, which it contains. It is a refrigerant, and an infusion or whey made by boiling it in milk, is a pleasant drink in febrile and inflammatory affections.

The Purselane *(Portalacca oleracea)* is a succulent plant, with small yellow flowers, and grows in cultivated grounds. It has an herbaceous, slightly saline taste, and is often used as greens, being boiled with meat or other vegetables.

The Yellow Dock (Rumex crispa) is a naturalized plant, originally derived from Europe. Its root is used medicinally. It is astringent, gently tonic, and is supposed to possess alterative properties, which render it useful in scorbutic disorders and cutaneous eruptions, particularly the itch. The powdered root is recommended as a tooth powder, especially where the gums are spongy.

The leaves of the Water pepper (Polygonum hydropiperoides) and of the Smart weed (Polygonum acre) have a sharp and biting taste, and are used as applications to ulcers, and are applied to the gums in mercurial salivation.

The Pokeweed (Phytolacca decandra) grows not only in waste places around fences, and in cultivated ground, but in the depth of the woods in the marsh lands in St. Mary's parish, where it reaches the height of from ten to fifteen feet. The young shoots are often used in early spring, and boiled in the manner of spinage. The berries contain a succulent pulp, and yield upon pressure a large quantity of a fine purplish red juice. They have a sweetish, nauseous, slightly acrid taste. The coloring principle is evanescent, and cannot be applied to useful purposes in dying, from the difficulty of fixing it. The taste of the dried root is slightly sweetish but is followed by a sense of acrimony. Its medicinal properties are emetic and somewhat narcotic. In small doses it is an alterative, and has been recommended in the treatment of chronic rheumatism. An ointment, prepared by mixing the roots or leaves with lard, has been used to advantage in some cutaneous diseases.

The Celery Crowfoot (Ranunculus sceleratus) is a naturalized plant, indigenous in Europe. It is pervaded by a volatile acid principle, which is dissipated by drying or by heat, and may be separated by distillation. The property for which it has attracted the attention of physicians is that of inflaming and vescicating the skin. It is a powerful rubefacient, far more efficacious than mustard.

The Virgin's Bower (Clematis Virginiana), the Leather Flower (Clematis Viorna), and the Crisp-flowered Clematis (Clematis crispa) are all ornamental vines, and well deserve a place in our gardens. The leaves and flowers have medicinal properties. They are useful as applications in cancerous and other foul ulcers, and in severe headaches.

The Dewberry (Rubus trivialis) and the high Blackberry (Rubus villosus) bear both very agreeable acid fruit, and are so abundant every where in Louisiana, that no one thinks of cultivating them. Their berries are much used as food, and a jelly made from them is in great esteem as an article of diet. Their root has tonic properties. Given in decoction, it is acceptable to the stomach, without being offensive to the taste, and it may be used advantageously in all cases where a vegetable astringent is of service, especially in children's complaints.

The Jersey tea (Ceanothus Americanus) is found every where in the United States. The root of this plant is astringent, and imparts a red color to water. The leaves were used during the revolutionary war as a substitute for tea, hence its name. A strong infusion of the dried seed and leaves is recommended as a local application to ulcers of the mouth and in sore throat of scarlet fever.

The Figwort (Scrophularia nodosa) is indigenous in Louisiana. The leaves have a bitter, somewhat acrid taste. They are said to be anodyne, and are sometimes employed in the form of ointment or of fomentation to painful tumors and ulcers and cutaneous eruptions.

The Mullein (Verbascum thapsus) is rather an unsightly plant, and is somewhat rare in Louisiana. I have only met with it near Ville Platte, in St. Landry parish, and seems to be introduced. The leaves and flowers have a narcotic smell, which, in the dried flowers, becomes agreeable. Their taste is mucilaginous, herbaceous and bitterish. They are emollient, and are said to possess anodyne properties, which renders them useful in chronic diseases. They impart their virtue to water by infusion.

The three-leafed Nightshade (Trillium sessile) is a pretty little herbaceous plant, and grows in the swamp lands of the Amite river. The roots are reputed to possess valuable remedial properties. They are employed by the Indians, and have been used by the country people. They are astringent and tonic.

The Common Nettle (Urtica dioica) is a well known plant, growing by the roadside and at the edge of gardens. This species of nettle produces, upon the slightest touch, a burning pain in the fingers, which continues for some time. Its irritant effect is said to be owing o the presence of free formic acid in the sharp and tubular hairs. The young shoots are boiled and eaten by some people as a remedy in scurvey, and the fresh plant is sometimes used to excite external vesication in cases of torpid and local palsey.

The Vervain (Verbena officinalis) grows most abundantly in every part of the State, near towns and cultivated fields. It has long spikes of small blue flowers, and blossoms from the beginning of summer till late in autumn. It was highly esteemed by the ancients both as a medicine and as a sacred plant employed in certain religious rites. In modern times superstitious notions in relation to its virtues are still entertained, and the suspension of the root around the neck by means of a white ribbon has been gravely recommended for the cure of scrofula. The leaves when bruised and made into a cataplasm, are used by country people as a remedy in severe headache and other local pains. Its real medicinal virtues are somewhat doubtful.

The Water Cress (Nasturtium officinale) has a fleshy stem, and grows in springs, rivulets and ponds. When fresh it has a quick, penetrating odor, especially when rubbed, and a bitterish, pungent taste. It is used in spring as a salad, and employed sometimes in scurvey.

The wild Pepper-Grass (Lepidium Virginicum) is a little plant which is widely diffused all over the United States. Its leaves, when chewed, have a pungent taste, and are much used by poor people as a salad in spring when other vegetables are scarce.

The Shepherd's Purse (Capsellia bursa-pastoris) is a species of the mustard family, and grows everywhere in this country. It yields a volatile oil, which may be obtained by distillation. The plant is bitter and pungent, and is supposed to possess astringent properties, and on this account it is employed in hemorrhages. The fresh herb when bruised is used as a topical remedy in rheumatism.

The Hedge Mustard (Sisymbrium officinale) grows along fences and walls. It has an herbaceous and acid taste, which is strongest in the tops of the flower-spikes and resembles that of mustard, but is much weaker. The seeds have considerable pungency. The herb has been recommended in chronic coughs and hoarseness, and ulceration of the mouth and throat.

The American Senna (Cassia Marylandica) is common in all parts of the United States. The leaves alone have medicinal properties. They have a feeble odor and a nauseous taste, analogous to that of senna. American senna is an efficient medicine, closely resembling the imported senna in its action, and capable of being substituted for it in all cases in which the latter is employed.

The Pepper mint (Mentha piperita) is a native of Great Britain, from whence it has been introduced into this country. It grows wild near Clinton, in East Feliciana. It has a peculiar, penetrating, grateful odor. The taste is aromatic, warm, pungent, glowing, camphorous, bitterish, and is attended with a sensation of coolness when the air is admitted into the mouth. These properties depend on the volatile oil which abounds in the herb, and may be separated by distillation with water. It is a grateful aromatic stimulant. It is used to allay nausea and relieve spasmodic pains.

The Horse mint (Monarda punctata) grows in light sandy soil all over Louisiana. The whole herb has medicinal properties; it has an aromatic odor and a very pungent, bitterish taste, and abounds in volatile oil, which may be separated by distillation. Its medical properties are stimulant.

The Common Nightshade (Solanum nigrum) is an herbaceous, mean looking plant, very widely diffused. The leaves alone are employed as a medicine. They have been used in cancerous, scrofulous and scorbutic diseases, being given internally, and applied at the same time to the parts affected in the form of poultice, ointment and decoction. Neither the berries nor the leaves are believed to be poisonous.

The Jerusalem Cherry (Solanum pseudo-capsicum) is a shrubby plant, bearing a red cherry-like fruit. I found it growing wild in East Feliciana. It is an ornamental plant, and is cultivated in the gardens on account of its bright green leaves and its red berries, which are said to be poisonous.

The Plantain (Plantago major) grows in fields, roadsides and grass plots. The leaves are saline, bitterish and austere to the taste. The root is saline and sweetish. The plant has been considered refrigerant and astringent. The ancients esteemed it highly, and used it in diseases where astringents are properly employed. The root is said to have been useful in intermittents. Among country people it is used as a dressing for blisters and sores.

The Thorn Apple (Datura stramonium) is supposed not to be indigenous in this country, but to have originated in South America or Asia. All over the United States the plant is generally known by the name of Jamestown Weed, a name derived, probably, from its having been first observed in the neighborhood of that old settlement in Virginia; and it is even a traditional story that the followers of John Smith had actually cooked it as greens, and had experienced its poisonous properties by eating of it. The leaves, as well as the seed, are considered medicinal It produces powerful narcotic properties. It has been administered with good effect in mania and epilepsy. It has also been found beneficial in neuralgia and rheumatic affections. It has acquired well deserved reputation in asthma. Smoking the leaves or seed during the paroxysm greatly alleviates, and often subverts it.

Fennel (Foeniculum vulgare), which, on account of its smell, its clusters of yellow flowers, is somewhat an ornamental plant in the gardens, has an aromatic odor and taste, dependent on a volalile oil by which it is pervaded, and which may be separated by distillation. Fennel seed was used by the ancients. It is one of the most grateful aromatics, and is employed to disguise the taste of other less pleasant medicines.

The Water Hemlock (Cicuta maculata) grows in wet grass lands and on borders of streams, to the height of from four to five feet. Its roots are very poisonous, and it has been recommended as a specific in nervous and sick headaches.

Boneset (Eupatorium perfoliatum) is a plant which is easily recognized by its small bushy white flowers, and by the long narrow leaves, which may be considered as being perforated by the stem. It abounds in moist and wet places. It flowers from the middle of August to the last of October. All parts of it have active properties. It is a tonic and diaphoretic. In large doses it is emetic. It is said to have been employed by the Indians in intermittent fevers, but its efficacy in that disease is doubtful.

The Yarrow (Achillea millefolia) is a perennial herb common to the old and the new continent. Both flowers and leaves have an agreeable though feeble aromatic odor, and a bitter, astringent and pungent taste. The aromatic property is strongest in the flowers, and the bitter principle in the leaf. It owes its virtues to a volatile oil, which may be obtained by distillation. Its medical properties are tonic, aromatic and astringent.

The May weed (Maruta cotula) grows abundantly in and around the town of Baton Rouge. It is undoubtedly introduced. Its flowers consist of a golden yellow disk and white radial florets. It is frequently called wild chamomile. The whole plant has a strong disagreeable smell and a warm bitter taste, and imparts its active principle to water. It may be substituted for chamomile, for its medical properties are the same. It has been given in nervous diseases, especially hysteria.

Canada Fleabane (Erigeron Canadensis) is a common weed which grows from two to six feet high. The leaves and flowers are said to possess peculiar virtue. It has an agreeable odor, and a bitterish acrid somewhat astringent taste. Both alcohol and water extract its virtues. It appears to be tonic and astringent, and has proved useful in dropsical and other complaints.

The Sweet scented Golden rod (Solidago odora) grows in the woods and fields in Louisiana. The leaves have a fragrant odor, and a warm, aromatic, agreeable taste. It is aromatic, and moderately stimulant and diaphoretic when given in warm infusion.

The Sunflower (Helianthus annuus) is considered, on account of the large size of its flowers, as an ornamental plant of the gardens.

It is a native of South America. The pith contains nitre, and is employed as a moxa in the cauterization of the skin by fire.

The Burdock (Lappa major) grows most luxuriantly in and around Baton Rouge, and a stranger might think it is cultivated on account of its medical properties. But its actual virtues are too insignificant to deserve much attention. The root has a weak mucilaginous and sweetish taste, and a slight degree of bitterness and astringency. It has been recommended in scorbutic, scrofula and rheumatic affections.

The Pinkroot (Spigelia Marylandica) is found in rich soil and on the borders of woods. The root, which is medicinally employed, consists of slender branching fibres, attached to a knotty head. It has a faint, peculiar smell, and a sweetish, slightly bitter, not very disagreeable taste. Its virtues are extracted by boiling water. It is considered as one of the most powerful anthelmintics. The Cherokee Indians were well acquainted with the vermifuge property of Spigelia, and the use of it, for that purpose, has been adopted from them.

HORTICULTURAL AND AGRICULTURAL PLANTS.

The native country of the Garden Lettuce (Lactuca sativa) is unknown. It has been cultivated from time immemorial. It is used as a salad, and when properly dressed, forms one of the most pleasant condiments. It abounds in a milky juice, which, during the flowering season, escapes readily by incision in the stem, and yields the lactucariun of commerce. It was well known to the ancients that lettuce possessed soporific properties. A tincture of lactucariun produces the anodyne effects of opium, without being followed by the same injurious effects. It is especially employed to allay cough and quiet nervous irritation.

The Tomato (Lycopersicum esculentum) is a well dnown vegetable served up at our table. It grows most luxuriantly in the gardens of this State, and bears fruit until the frost kills the vine. It is of a bright red color, and on this account has been called the love apple. The fruit has a peculiar acid taste, and is highly nutritive. It is said to possess anti-bilious properties.

Cayenne Pepper (Capsicum annuum) is a native of the East Indies and tropical America, and is extensively produced in this country, both for culinary and medicinal purposes. It is a powerful stimulant, producing, when swallowed, a sense of heat in the stomach and a general glow over the body without narcotic effects. It is much employed as a condiment, and proves useful in aiding digestion. In India it has been used from time immemorial, and from a passage in Pliny it appears that it has been known to the Romans. As a medicine, it is useful in cases of enfeebled and languid stomach, and is occasionally prescribed in dyspepsia. Applied externally, it is an excellent rubefacient in local rheumatism. It acts speedily without producing vesication.

The Ground Nut (Arachis hypogœa) is indigenous in Africa and South America. It is cultivated on a large scale near Wilmington, North Carolina. A remarkable property of the peanut is that its pods penetrate the soil in the progress of their growth, and that the fruit ripens under the surface of the ground. The seeds constitute the well known ground nuts in our markets. These when roasted form an agreeable article of food. A fixed oil is expressed from them which is used for lubricating purposes, and which renders it a valuable article of commerce.

The Bean (Phaseolus lunatus and P. nanus) is cultivated in the garden as well as in the field for the use of its large nutritious seed. The bean is mentioned in all the ancient classical works in which reference to agriculture is made. It was esteemed more than any other kind of leguminous vegetable, both by the ancient Greeks and early Romans. The Athenians used sodden beans in their religious festivals in honor of Apollo, and the Romans presented beans as a sacrificial offering in honor of Carna the wife of Janus. In Egypt it appears to have been regarded as typical of some of the mysteries which the priests endeavored to conceal from the knowledge of the uninitiated, and it was therefore shunned by them as too sacred for ordinary observation. Pythagoras, who studied philosophy on the banks of the Nile, and lived exclusively on vegetable food, forbade the use of the bean. Beans were employed by the Romans in taking the vote of the people and they used them in the election of magistrates. The meal or flour made of them was thought to possess the property of removing wrinkles and giving a fair complexion to the skin.

The bean is indigenous in Egypt, Barbary and Morocco. The lima bean and the string bean are the principal varieties cultivated. A rich, strong loam is best for their successful cultivation, and if properly prepared, will produce a crop of fifty or even a hundred per cent. The penetrating, ramifying, fibrous root of this vegetable cleaves and subdivides the stiff soil, so as to draw down a free circulation of the air, and to dry, pulverize and mellow the loamy earth. Its large leaves absorb a great quantity of nourishment, in the form of atmospheric gases, and by their fall communicate the elements collected to the soil. A crop of beans is the most proper means for improving wet and heavy land; while eliminating nearly as much nutritious matter for the use of animals as a crop of wheat, it produces a far less exhausting effect upon the soil, and upon heavy land it excels every other crop in making a remunerating return for manure. They are better suited for feeding horses and are more nutritious than oats. When thus used, they ought to be split or bruised in a mill and given in a mixture with cut hay or straw. They are also much employed in England for the fattening of hogs, but they have a tendency to render pork firm, and not sufficiently delicate, and at the last stage of fattening they are superseded by barley meal. Bean meal is well adapted to fatten oxen; and mixed with the drink of cows it very materially increases the yield of milk. Wheat flour is often adulterated by bean meal.

Cabbage (Brassica oleracea) has been generally and very early cultivated by the ancient Greeks and Romans. The Greeks held it in great esteem, and are said to have expatriated their physicians; and simply by the use of cabbage have preserved their health for six hundred years. Both Greek and Romans ate the raw leaf as a preventive of intoxication, and as alleviating its effects. Pliny, speaking of the spring shoots, says: "I dwell long on this vegetable, because it is in so great repute in the kitchen and among our riotous gluttons."

The most suitable soil for cabbage is a sound mellow loam, of the peculiar quality and texture designated as fat or unctuous, containing sand in very large proportion in minute division, combined with a considerable quantity of aluminous earth. When cabbage is fed to milch cows, all the unsound leaves should be carefully stripped off. This vegetable more than any other plant contains nitrogenous materials, suitable for the nourishment of man, and is on this account a good substitute for meat.

The Turnip (Brassica rapa) is a biennial and has two seasons of growth, one in which it develops its leaves directly from the root crown, and another in which it sends up a long flower stem with leaves entirely different from those of the first season, and then goes through the process of flowering and seeding. The turnip has been introduced from Europe by the early settlers.

The Sweet clover (Melilotus officinalis) is indigenous in Europe, and has been introduced into this country. The flower has a peculiar sweet odor, which by drying becomes stronger and more agreeable. Its taste is slightly bitterish. It is tender and succulent when young, but becomes hard and woody in the stem, and is then only fit to be eaten in the tops and leaves, when in full flower. It is greedily eaten in its young and tender state by all kinds of live stock, and when dried into hay has a fragrant odor. On account of its flavor it is used in making the famous Gruyere cheese. In medicine it is employed in the form of cataplasm applied to slight inflammation.

The Red clover (Trifolium pratense), besides the excellent grazing it affords to cattle and horses, covers the ground with its broad foliage so as to smother useless annual weeds. It enriches the soil by the fixation of gases, and by the profuse ramifications of its roots it acts with the power of a fallow, and makes both a mechanical and chemical preparation for a beautiful and luxuriant cereal crop. It flourishes best in limestone soil. There are biennial and perennial varieties.

The White Clover (Trifolium repens) is an indigenous perennial. It grows wild, and possesses such extraordinary vitality as to spring up spontaneously in great profusion in places where it could not have previously vegetated for centuries. Its roots are fibrous, its stems are stoloniferous and creep along the ground, striking root as soon as they touch it.

The Procumbent Hop (Trifolium procumbens) has yellow flowers, and grows wild in East Baton Rouge. It is probably introduced. It is but little relished by any kind of live stock, and is exceedingly liable to mildew.

Timoty (Phleum pratense) has been introduced into this country from Europe, where it is indigenous. It thrives best in moist soils and rich alluvial clay land. It affords twice as much nourishment when its seeds are ripe, which takes place in August, as when it is cut while in flower. It is valuable as a permanent grass and for making hay, as well as for alternate husbandry, to be plowed under to prepare the ground for another crop.

Common Oats (Avena sativa) have been cultivated in Europe from time immemorial. The oat grows but moderately in the colder parts of the world, and flourishes best in the middle regions of the temperate zone. It becomes sickly and unproductive on approaching the tropics. Oats are easily raised on almost any kind of soil, from the heaviest loam to the lightest sand.

Indian Corn (Zea Mays) is indigenous in the West India Islands, where it reaches the height of from twelve to fourteen feet, and completes its course of vegetation in forty days. It flourishes between the fortieth degree of south latitude and the forty-fifth degree of north latitude. It is extensively produced in some parts of Asia and Africa. It is the common bread corn of the Levant, and of a large part of Spain, Italy and southern France.

It has made its way even to many parts of Germany, Belgium, and the southern and central parts of England. Its widest range of production extends over the whole of the Middle, Western and Southern States, as well as Mexico, where it constitutes one of the principal cereals for making bread, and the chief material for feeding cattle and horses.

Sugar cane (Saccharinum officinarum), or the material derived from it, is one of the principal staple products of Louisiana plantations. It is mostly cultivated in the coast and gulf parishes. It is a perennial plant belonging to the family of grasses. It varies in height from six to fifteen feet. The stalk is knotty, with a leaf and inner point at each knot. It is propagated by slips or pieces of stem with buds on them. There are three kinds of cane which are cultivated for the production of sugar. The old creole cane, which has dark green leaves, a slender stem, and is close jointed. It came originally from India, and was introduced in the West India Islands. The cane from Otaheite is the most valuable. It has light green leaves with a thick stem, grows to a considerable height, is very juicy, and presents a luxuriant appearance. The cane from Batavia is indigenous in Java. Its purplish leaves are very long and broad, and the juice which it produces is preferred for the manufacture of rum. The sugar cane has been introduced into Louisiana from the French West India Islands.

Sugar was known to the ancients. Dioscorides refers to the canes of India and Arabia Felix as producing honey; and Pliny speaks of the saccharum of Dioscorides as being used in medicine.

Sugar lands, unless naturally rich, require an abundance of manure; but saline matter should be avoided, since common salt (chloride of sodium) and muriate of ammonia form non-crystalizable compounds with sugar. Within the crisp rind of the cane is a whitish, porous pith, saturated with saccharine matter. The juice is expressed by passing the cane through revolving cylinders. From the mill it runs into large boilers, in which it is heated for purification. Lime is used in sugar boiling for the purpose of neutralizing the free acetic acid which exists ready formed in the woody parts of the cane, and to clear it from various foreign materials mingled with it. By application of gradual heat, the impurities which sugar contains form a cake with the lime at the surface of the saccharine liquid, which is drawn off and conveyed to the boilers. After having passed through several boilers, it becomes a dark thick syrup, when it is put into flat coolers to crystalize.

The crushed cane, known as begasse, is used as fuel in evaporating the juice, but it would be far more preferable to return it to the land as manure.

Cane juice is a solution of sugar in water, with traces of albumen, gum, and a peculiar substance resembling gluten, or vegetable gelatine ; also a minute proportion of cerasin and of a green vegetable wax. It has usually a yellowish color, but is sometimes colorless. It has an agreeable, but rather insipid taste, and a peculiar balsamic odor. It contains from seventeen to twenty per cent. of cane sugar, but the planter only obtains from seven to ten per cent. There is a loss of sugar in the mode of pressing the juice from the cane, and a loss from chemical change due to the exposure to the air, by which the crystallizable sugar becomes degraded into non-crystallizable mucilaginous sugar, called molasses.

Raw sugar is refined by mixing with it a small portion of lime water, bullock's blood, and a quantity of animal charcoal.

Louisiana produced, in 1860, about five hundred millions of pounds of sugar, at the value of twenty-five millions of dollars ; and thirty-five millions of gallons of molasses, at the approximate value of seven millions of dollars. It has been estimated that before the war Louisiana produced an average crop of four hundred and forty-nine thousand hogsheads, or two hundred and twenty-five tons of sugar, which supplied forty per cent. of all the sugar consumed in the United States, amounting, in 1865, to four hundred and three thousand one hundred and nine tons. It is believed by men conversant with the subject that during this crop season there will be one thousand one hundred and seventeen sugar houses in operation, against one thousand two hundred and ninety one at the commencement of the war, and eight hundred and seventeen in 1869.

Good sugar lands will produce from a thousand to fifteen hundred pounds of sugar and seventy gallons of molasses to the acre. Under the present system of labor, one hand can cultivate, besides the other crop, ten acres of cane, which, at ten cents for sugar and sixty cents for molasses, produces the handsome sum of one thousand three hundred and thirty-six dollars Sugar growing is, therefore, one of the most profitable agricultural operations, and is alone sufficient to make Louisiana one of the richest States of the Union, provided the money were not, in great part, carried away by Northern merchants and speculators, who make fortnnes by Southern commerce, and build palaces in New York, where they and their families live in magnificence and splendor.

Cotton (Gossypium herbaceum) is one of the principal agricultural products of Louisiana, second in importance only to sugar. It is an article of commerce of such great value that it places New Orleans in the first rank of the exporting cities of the world. Herbaceous cotton is chiefly cultivated in the Southern States and the East Indies. It grows from two to eight feet in height, is rich in foliage, and its fibrous seeds are preceded by flowers of pale yellow and crimson color, like that of the hibiscus, for it belongs to the mallow family. As the flowers fall, a capsule is found containing the fibres of cotton which constitute the covering that envelop the seeds. Cotton was cultivated in the East Indies five centuries before the Christian era. The clothing of the Hindoos consisted chiefly of garments made of this vegetable product. But the India cotton is of shorter staple, and consequently much inferior to that of the Southern States. In Borneo and many of the tropical regions, the plant on which the cotton grows is flourishing in a wild state. In the West Indies, in Brazil and in Egypt it is produced from a shrub.

There is also a variety of herbaceous cotton called Sea Is and cotton, which grows nowhere else except in Florida and the islands on the Georgia and South Carolina coast. It is very valuable on account of its long fibre, and is only employed in the manufacture of lace and other costly articles of luxury.

Cotton fibre, when examined by the microscope, is found to be somewhat flat, and bluntly triangular. Its direction is not straight, but contorted so that the locks can be extended and drawn out without doing violence to the fibres. The threads are finely toothed, which is the reason for their adhering together.

The New Orleans and Mobile cotton is most valued in the European markets. Before the war four millions of bales, at four hundred pounds a bale, were produced, the value of which was estimated at one hundred and fifty millions of dollars. Louisiana produced before the war five hundred and fifty thousand bales of cotton, whose value was no less than twenty-two millions of dollars.

Cotton seed contains a fixed oil which is expressed and sold for lubricating purposes. It is valuable as food for cattle, and as manure for exhausted cotton lands.

Rice (Oryza sativa) was cultivated in the East from time immemorial. It is the most wide spread of all the cereal grasses. In Southern Asia it is almost the only food of the lower classes. The Japanese and Chinese, and the people of the East Indies, of Madagascar, Persia and North Africa would be exposed to great suffering without a sufficient supply of rice. It flourishes in the countries north of the Mediterranean—Turkey, Greece, Italy and Spain. It has been found in a wild state in the interior of South America. It forms the principal staple product of the coast plantations of Georgia and the Carolinas, and it is cultivated to a limited extent on some of the coast plantations below New Orleans. It is an annual grass growing up with a stalk similar to that of wheat, having the joints, however, much closer, and much more numerous. The grains are enveloped in rough yellow husks, provided with an awn. The seeds divested of the husks constitute the rice of commerce.

There are several varieties. The common rice grows only in marshy soil, and requires irrigation at certain stages of its growth. The mountain rice (Oryza mutica) is cultivated in Cochin China and Java, and flourishes best on the slopes of hills.

Tobacco (Nicotiana tabaccum) is probably a native of tropical America. It is now cultivated in most parts of the world, of which the Cuba and the Virginia tobacco are the most celebrated. Tobacco acquires a strong penetrating odor after it has undergone the necessary manipulation to prepare it for market. The fresh leaf is without odor. It has a bitter, nauseous and acrid taste. These properties are imparted to water and alcohol. Its active principle which has been separated has received the name of nicotia. In its action on the animal system it is the most virulent poison known. A drop of it in the state of concentrated solution is sufficient to destroy a dog; and small birds perish at the approach of a tube containing it. In man, it is said to destroy life in poisonous doses in from two to five minutes. Tannin might be employed as a counter-poison.

Tobacco unites the powers of a sedative and narcotic to that of an emetic. It produces this effect to a greater or less degree to whatever surface it may be applied. When chewed it irritates the mucous membrane of the mouth and increases the flow of saliva. Moderately used, it quiets restlessness, calms mental and corporeal inquietude.

The use of tobacco was adopted by the Spaniards from the American Indians, some tribes considering smoking a religious ceremony. Nuttall tells us that an Indian chief informed him that the Osages smoked to the Great Spirit or to the sun, and accompanied it by the following apostrophe : "Great Spirit deign to smoke with me as a friend; fire and earth smoke with me and assist me to destroy my enemies; my dogs and horses smoke also with me."

Tobacco was introduced into France by the French ambassador at the court of Portugal, whose name, which was Nicot, has been immortalized in the generic name of the plant. Sir Walter Raleigh is said to have introduced the practice of smoking into England.

Smoking, when indulged in to excess, enfeebles digestion, produces emaciation and general debility, and lays the foundation of serious nervous disorders. Four cases of insanity have been reported to have occurred in the Pennsylvania hospital of the insane, the cause of which had been ascribed to the abuse of tobacco. Its violent action as a remedial agent prevents its frequent employment in medical practice, and it is chiefly used as a narcotic to produce relaxation in spasmodic affections.

The spectroscope analysis has shown that tobacco ash contains the rare mineral called lithia, which imparts such a beautiful red color to the flame.

Tobacco is not much cultivated in Louisiana for general consumption. The perique is, perhaps, the only kind which has acquired considerable celebrity, for its strength and the purity of its materials, and is preferred by inveterate smokers to every other kind of the manufactured article. It is principally cultivated in St. James parish, and perhaps, also, in Natchitoches.

MICROSCOPICAL BOTANY.

This new branch of botanical science, brought to light by Professor Ehrenberg, of Berlin, who was the first to dicover fossil silicious shells, and extending his investigations, he found many living types exactly identical with those in a fossil state. As these little organisms move freely in water, it was at first supposed that they belong to the animal kingdom; but it is now established beyond all controversy that the desmidiae and diatomaceae are unicellular plants belonging to the class of fresh water and marine algae. Since that time an extensive microscopic flora has been discovered, and a great number of specimens have been described, existing in every part of the world.

The diatomaceous algae present the most beautiful forms, the most exquisite carvings, and the most brilliant coloring. This branch of botany possesses as much fascination, and excites an interest as intense as that felt by the astronomical observer, who watches with close attention, with the aid of his instrument, the course of the stars and the revolutions of the planets, plunging into the world of the infinite and the unknown; thus unveiling the secrets of nature, and extending his ravished vision beyond the ken of ordinary mortals.

The student of microscopic botany transfers himself into a new world; not to that which is infinitely great in space; not to that which, though of immense magnitude, can not be seen by unaided vision, on account of the vast distance which intervenes between it and the eye of the observer; but to that world which is near and within reach of human sight, but is nevertheless invisible to the human eye, on account of its almost inconceivable diminutiveness. The most minute diatom, which in size measures only one-thousandth part of an inch, is as much the work of the Creator as those immense masses of worlds, in comparison to which our earth is a mere speck in the immensity of space. Its general outline and beautiful symmetry of form, as well as its mode of life and manner of propagation, is as wonderful as the most astounding astronomical phenomena. It is a new world, which, so to say, has been called into existence during this nineteeth century, the age of great discoveries, a world diversified by hundred different forms, peopled by hundred different, living, moving, self-producing organisms, existing in a single drop of water. I cannot resist the temptation of adding here the beautiful motto of Greville's Scottish Cryptogamic Flora, in the original Latin, as it would loose much of its beauty by translating it: "*Cui bono haec omnia? Ut cognoscamus sapientium Creatoris, quae in minimis non minus educat quam in magnis plantis.*"

Prompted by the desire of studying this branch of science, which ought to be cultivated in every school of a high grade, I undertook, with the University microscope at my disposal, the most attractive study the scientific explorer can be engaged in—the examination of the Microscopic Algæ of Louisiana. Unfortunately the magnifying power of the various object glasses of our microscope is not exactly known. It may however be safely assumed that its highest power ranges from between 200–250 diameters. I have been enabled by means of this instrument to determine from eighty to ninety species of Desmidiæ and Diatomaciæ, most of them with complete certainty; a few only approximately, taking Pritchard's Infusoria as my guide, and consulting Rabenhorst's "Flora Europæa Algarum."

Most of the specimens, of which a list is annexed, have been microscopically viewed while in a living state, and have been carefully studied after preparing and mounting them in Canada balsam. They are thus permanently fixed, and their place on the slide, which is precisely indicated by means of that admirable and simple little contrivance, called Maltwood's Finder, is registered in a book kept for that purpose, so that the specimens can be examined at pleasure at any future time. I intended to accompany this report with the figures of all the microscopic plants determined, but the appliances required to accomplish the object have not yet been received, and consequently that part of my report will not have as much interest, as far as men of science are concerned, as it would otherwise possess, if it had been possible to give the illustrations.

As there is no microscopist in the United States, who, as far as I know, has taken the place of the late Professor Bailey, of the West Point Military Academy, it would be well that the Louisiana State University should adopt this branch of natural science as a specialty, and thus contribute by scientific research to the diffusion of scientific knowledge, in which the scientific world takes considerable interest. It must not be understood, however, that I have the least pretension of being able to fill the place left vacant by Professor Bailey. I can only promise, that I shall use my humble abilities to the utmost, in order to give a more extensive range to microscopic investigations as regards the Algæ of Louisiana and the Southern States.

But to accomplish this object successfully and give authority to my determinations it would be necessary to supply all the means which the modern improvements of the microscope afford. It is some months, since you have ordered Ehrenberg's Infusoria and Ralf's British Desmidiæ to be purchased for the use of the University, but they have not yet been received, and consequently, I could not avail myself of the information these works would have furnished me to render my report on microscopic plants more complete. It would also be of great advantage, and in some cases it would be indispensably necessary, to possess prepared slides, containing specimens of all the diatomaceæ thus far found in the United States, for comparison, so as to determine new species with certainty. Additional object glasses ranging from three hundred to a thousand diameters would also be necessary, for some of the markings of the diatomaceæ are so delicate that they can only be seen with object glasses of very high powers, and the determination frequently depends on these markings.

A great number of these microscopic plants, many of which are probably new species, are still undetermined, not only for the reasons already stated, but, also, because I have not been able to correspond with scientific men in this country who are familiar with this subject, and I must, therefore, defer to a more propitious time to furnish the full catalogue of the specimens found and preserved for future examination.

THE FLORA OF LOUISIANA.

In undertaking the botanical survey of the State and collecting every species of plant that grows in Louisiana, my object has been to make the collection valuable, not only for the purpose of obtaining a succinct view of the inexhaustless resources of wealth derived from the vegetable kingdom, scattered over the State, but to offer to the young student of botany a scientific collection of authentic specimens, which he may consult whenever any difficulty may present itself in the determinination of plants, he may meet with in his botanical excursions.

When the list of plants, constituting the Flora of Louisiana, shall be completed, it will afford considerable interest to the scientific botanist, not only because it includes some few new species, but more especially, because it furnishes him with new data to determine the range and geograpical distribution of plants. This is a subject which has not yet received, on account of the insufficiency of information, the attention its importance seems to merit.

To determine and classify nearly a thousand specimens of flower-bearing plants, of mosses, lichens, ferns, algae and fungi, is a task of considerable magnitude. A botanist who wishes to study the stages of growth and development, as well as the specific characteristics of the numerous individual plants embraced in the various branches of botanical science, needs not only the assistance of an extensive library, consisting of books written in different modern languages, as well as Latin, but he must, in addition, have access to collections of authentic specimens, placed at his disposal for consultation and comparison. With all these advantages he would nevertheless encounter many difficulties, and meet with anomalies which he would be unable to explain, without advising with scientific men of long experience and established reputation. Although additional books on botany have been supplied since my last report was written, yet want of time, and the insufficiency of standard works on the cryptogamous plants of the Southern States, have compelled me in many instances to apply to the scientific botanists of the United States to lend me the aid of their experience in the determination of many of the grasses and sedges, and a great number of the non-flower bearing plants. I have corresponded with scientific men in various parts of the country, and have everywhere received the most cordial recognition. I consider it, therefore, a pleasing duty to acknowledge the kind assistance received from Dr. John Torrey, L. L. D., of the city of New York, one of the most eminent botanists of this continent, in the determination of many of the grasses and sedges. I am under great obligations to C. F. Austin, Esq., of Closter, New Jersey, a bryologist of great experience, for his invaluable services in the determination of the mosses and lichens. I also return my best thanks to Rev. M. A. Curtis, D. D., of Hillsborough, North Carolina, one of the ablest mycologists in this country, who has kindly determined for me many of the fungi thus far collected.

I am indebted to the Smithsonian Institution, through the agency of Prof. Joseph Henry, L. L. D., its able Secretary, for the determination of a portion of the shells collected at Grand Isle and elsewhere, of which a list is annexed.

The books consulted in the determination of plants, and with reference to their medicinal and economical use, are: Gray's Manual of Botany of the Northern United States, Dr. Chapman's Southern Flora, Dabney's Botany of the Southern States, Brown's Trees of North America, Wood & Bache's Dispensatory of the United States, Bruch & Schimper's Bryologia Europæa, Pritchard's Infusoria, Rabenhorst's Flora Europaea Algarum, and the Rural Cyclopaedia.

In the list of plants of the Flora of Louisiana, I deemed it of some importance, in order to render the study of botany more popular, to annex to the botanical name the English name, if such has been in common use; but in the cases of plants which have not yet received an English designation, I anglicised the generic botanical name, and translated the specific name, as far as this was practicable, so as to correspond with some of the characteristics of the plant which it was intended to designate. I have also given the locality where each species of plant was collected, that the report might acquire additional interest by attaching to each specimen some local association. To avoid misapprehension, it is perhaps proper to say that it must not be supposed, because a certain locality is indicated, that the plant grows no where else in the State; on the contrary, most plants have a very extensive range of growth, and there are only very few which are found in some favored spots, and are met with no where else, within certain circumscribed limits.

Some few specimens gathered in the parishes west of the Mississippi, are believed to possess sufficient diversity of characteristics to consider them as new species. They are fully described, so as to enable scientific botanists to recognize them by their specific difference, or refer them to a species previously described and named by other authors.

I have taken the liberty, without previous consultation, to name a new species of Jussiæa, "Jussiæa Boydiana," as a well merited compliment to you, the superintendent of the University, for your valuable services, and for the efforts made by you in behalf of the botanical survey of the State.

CONCLUSION.

Although the list of plants collected within the last twelve months is quite extensive, yet it must not be supposed that, within such a short period of time, I could have collected the greatest number of plants which grow in the State of Louisiana. With the exception of eleven new specimens contributed by R. S. Jackson, Esq., for which he has our thanks, the whole collection has been made by myself personally, and the labor of manipulating the specimens, to prepare them for preservation, has exclusively devolved upon me. Even if the collector could be at different places at the same time, which the accomplishment of such a work, within a given period, would require, it would be impossible to dry and prepare, determine and classify all the plants of Louisiana in a single year. We have comparatively few of the spring and late fall flowers, and these are the seasons when a great number of plants produce their blossoms, and though the number collected will not be much less than one thousand specimens, which have been classified and determined, yet this probably constitutes not more than one-third or one-fourth of the aggregate vegetable productions that grow on Louisiana soil or in Louisiana waters. It would require many years to visit the various localities where the cryptogamous plants flourish best, and, in a scientific point of view, they present a far higher interest than the flower-bearing plants. The trees of Louisiana are of the highest importance, and a whole year might be profitably devoted to their study. A description of the productions of the forest, with the history which each species of tree has in the traditions of the aborigines, who were botanists in their way, as well as the traditions of the early settlers and country people in connection with particular trees, a systematic investigation of the quality of the wood each kind of tree is composed of, its mechanical, economic and artistic uses, would furnish the most interesting as well as the most useful compilation the botanist could engage in. To accomplish this object, assistance would be needed, information would be required from the most intelligent class in each parish, especially physicians, who travel about in the country, and whose profession leads them to botanical research. The University might by this means obtain a cabinet of sections of wood of all the forest trees and shrubs which grow in Louisiana. These sections should be in duplicate, one in its natural state, and the other smoothed and highly polished by the skill of the artist. The quality of each species of wood, unless already known and described in other works, should be ascertained by experiment, aud the University would thus become the depository, not only of useful information, but of samples of the vast economical resources of the forests of Louisiana.

During this session, botany has been introduced as a branch of study in the University, and a course of lectures on that subject has been commenced, and a small class, composed of zealous and attentive students, has been organized.

More time is required to accomplish successfully the object of the botanical survey. One month of the spring season, from the fifteenth of April to the fifteenth of May, ought to be devoted to botanical excursions, in order to collect as many spring plants as possible in different parts of the State.

To enable the botanist to take a wider range and pass through a greater extent of country, it will be indispensably necessary to be provided with a Jersey wagon or a vehicle of that kind, to carry presses and drying papers along with him so as not to be restricted to a central locality as a point of easy communication. Excursions of this kind can not be made on horseback except for a short distance, and in most country neighborhoods I was compelled, because no horse could be procured either for love or money, to travel on foot for miles in the hot summer sun in order to reach localities I desired to visit.

The result of my labors are thus laid before you, and you are able to judge how the duties devolved upon me have been performed.

FLORA LUDOVICIANÆ.

DESCRIPTION OF THE NEW SPECIES REFERRED TO IN THE FOLLOWING LISTS.

1. *Euphorbia Ludoviciana.*—Stem erect, very slender, smooth throughout, alternately branched. Stipules broad lanceolate. Leaves very thin, oblong abovate, slightly tapering towards the base, short petioled, smooth entire. Flowers axillary. Involucre short-pediceled, nearly sessile. Glands six, with six whitish persistent appendages, oblong, obtuse. Capsule orbicular, slightly convex, somewhat flat-topped. Seed broad, rounded on one side and sharply angular on the other, so as to form a triangular solid, with the curved upper side, thickly sprinkled with minute blackish dots. This Euphorbia grows from one to two feet high, has an extremely slender, virgate stem. The branches are all simple, are long below and continue gradually to decrease in length to the top, where they give place to almost sessile leaves, with flowers in their axils.

Habitat.—It is found in St. Landry parish, in red, loamy soil.

2. *Euphorbia Meganæsos.*—Smooth throughout, stem very slender, ascending, bearing the short flower-branches. Leaves opposite, short-petioled, linear oblong, narrow, obtuse, oblique at the base, obscure dentate, transparently dotted, with a crimson streak along the midrib, the whole plant turning red when in seed. Stipules fringed. Flowers in lateral and terminal clusters. Appendages of glands white or rose colored. Capsule obtuse-angled rugose. Seeds reddish, in the form of a coffeebean, with a groove in the flat base dividing the narrower apex, and the slit extending to one-half of the convex surface.

Habitat—This plant grows in tufts, in great abundance, near the sand beach of Grand Isle, where nothing else but a few grasses and the Ipomæa pes capra flourishes. Its short stem and long branches are from one-half to one foot long.

3. *Sabbatia Nana.*—Stem simple, low, somewhat angled. Leaves small, sessile, the lower spatulate lanceolate, the upper linear lanceolate. Lobes of the corolla five to six, one-third longer than the

narrow linear calyx lobes. Root perpendicular, slender. Flowers rose-colored. Stem three to four inches high. Blooms in August.

Habitat—Marshy soil of Grand Isle.

4. *Sabbatia Oligophylla.*—Stem erect, somewhat four-angled, simple and scape like. Leaves opposite, long linear, lanceolate, acute, clasping, remote, the stem bearing but two pairs of leaves. Radical leaves nine, clustered, obovate, narrowed at the base, and spatulate lanceolate sessile. Flower solitary, terminal. Corolla large, rose-colored, nine parted, three times as long as the partially reflexed linear calyx lobes, four bracted, two of the bracts nearly as long as the petals, and the other two twice as long. Stem eight inches high.

Habitat—Grows in the pine and oak woods beyond Chicotville, St. Landry, in sandy soil, and blooms in August.

This species differs from Sabbatia-choroides, in having the stem angled and the leaves clasping. It has but two pairs of cauline leaves, and the others are all radical and clustered, and the flowers are bracted.

It differs from S. Boykinii in having four instead of two bracts, and the calyx lobes are not lanceolate, but narrow, linear and reflexed.

The S. gentianoides, which it seems to resemble, has no bracts.

5, *Hydrolea Leptocaulis.*—Stem spiny, ascending, smooth, simple or branched. Leaves varying from oblong lanceolate to lanceolate from three to five inches long, acute, smooth, glossy and shining, tapering into a petiole. Flowers large, axillary, mostly solitary, sometimes two in a cluster, pale blue, short-peduncled, almost sessile. Corolla five parted, divisions oblong obtuse with five triangular white spots at the base. Calyx lobes ovate lanceolate, smooth, two-thirds as long as corolla lobes. Stamens and styles exerted.

Habitat—Is found in the bayou near Washington, St. Landry. The outside appearance of this species is so much different from the H. quadrivalvis, that I considered it advisable to make a new species of it, though it may only be considered a variety of the latter. The leaves are larger, more deep green and glossy, the flowers are much larger and have a brighter color, besides being distinguished by the white spots at the base of the petals. The calyx lobes are not linear, but ovate lanceolate.

6. *Hydrolea Ludoviciana.*—Stem erect, simple or branched, fis-

tulous, sufrutescent, slender, terrete, hoary-tomentose, spiny, spines long and slender. Leaves ovate or ovate-lanceolate, tapering at both ends, acute or accuminate, entire, velvety-pubescent, with undulate margin, short petioled. Flowers corymbose, terminal, large, crowded, azure blue. Stamens and styles exerted. Styles two to three times as long as stamens, curved in opposite directions. Penduncles crowded with bract-like leaflets. Capsule globular smooth.

Habitat—Marsh prairies near Bayou Portage, St. Mary and near the prairie lakes, St. Landry. It flowers in August.

7. *Jussiaea Boydiana.*—Stem simple, smooth below, slightly pubescent above, ascending from a creeping base. Leaves spatulate lanceolate, tapering into a petiole, obtuse. Flowers large, calyx-lobes five, one-third shorter than the petals, lanceolate acute. Capsule linear, cylindrical, half as long as the pedicel. Stem one foot high. Flowers in August ; flowers yellow.

Habitat—This species of jussiæa grows in the Mississippi swamp near Port Hudson. It differs from J. leptocarpa, which it resembles, by its simple and almost smooth stem, its spatulate leaves, its larger flowers, and especially by its capsule being only half as long as the pedicel, while in the J. leptocarpa the capsule is twice as long as the pedicel.

It is named in honor of Colonel Boyd, Superintendent of the Louisiana State University.

8. *Tephrosia angustifolia.*—Stem slender, angled, pubescent, with appressed hairs, decumbent. Leaves short, petioled ; petiole rusty villous. Leaflets eighteen to twenty-five, cuneate, lanceolate, narrow, rounded at the apex, strongly mucronate, smooth, with prominent veins on both sides, hairy on the margin and midrib. Racemes slender, two or three times as long as the leaves; from fourteen to fifteen flowered. Calyx teeth, terminating in a long hispid point. Legume falcate, narrow, smooth. Flowers light purple.

Habitat—Pine barrens near Pontchatoula. Flowers in July.

9. *Tephrosia multiflora.*—Stem rusty pubescent, branched, decumbent; leaflets eighteen to twenty-six, nearly sessile, the short petioles woolly pubescent, mostly linear oblong or cuneate oblong, rounded or emarginate at the apex, strongly mucronate, smoothish above, long hairy beneath and on the margin. Raceme stout, two or three times as long as the leaves; eighteen to twenty-five flowered;

calyx hairy; calyx teeth, terminating in a villous point. Corolla hairy. Flowers purple, mixed with yellow. Legume smooth, eight to nine seeded.

Habitat—Pine barrens of Pontchatoula. Flowers in July.

10. *Lilium Lockettii*—Stem about one and a half feet high, naked, leaves radical, clustered, about seven inches long, one inch broad, linear lanceolate, remotely and obscurely dentate, many nerved with transverse interstitial nerves, so as to give the leaf the appearance of laticework. Leaves of the perianth laticed like the stem leaves, narrow, lanceolate, accuminate, white, four inches long. Tube of the perianth two inches long, narrow, attached to a two inch long peduncle. Flowers from three.to five, clustered around the summit of the stem.

Habitat—This lily grows in the Calcasieu prairies on the banks of Lake St. Charles, where Colonel Lockett, who conducts the topographical survey, and after whom it is named, first discovered it.

11. *Oenothera paludosa*—Stem slender, simple or branched, somewhat wing-angled, shrubby at the base, fistulous, covered with a scurfy minute pubescence. Leaves entire, narrow lanceolate, acute three nerved, tapering into a short petiole orsessile, one and a half inches long. Pod four-angled, not winged longer than the pedicel. Sepals lanceolate, terminating in a subulate point.

Habitat—This Oenothera grows in the swampy prairie in St. Landry, between Ville Platte and Chicotville. Its stem is from one to three feet high. It flowers in August. Flowers yellow.

12. *Helenium Seminariense*—Leaves oblong lanceolate, entire, obtuse, decurrent. Scales of involucre linear lanceolate. Ray florets three lobed, broad, cuneate, longer than the disk. Disk florets five-cleft. Disk globose. Scales of pappus ovate, long, awn-pointed. Stem, leaves and peduncles sparingly glandular hairy. Achenia smooth with two ciliate lines.

Habitat—Pine barrens, Rapides. Stem one and a half feet high. Flowers yellow and flowers in June. The flowers of this species are nearly twice as large as those of H. quadridentatum.

13. *Lippia nodiflora Mich., var. microphylla.*—Stem creeping, peduncles naked, rising from the creeping stem, four to five times as long as the leaves. Leaves fleshy, small, one-half to three-fourths of an inch long, three lines wide; spatulate lanceolate, tapering into a petiole, serrate above, entire below the middle.

Habitat—In marshy soil, near New Orleans and Brashear City. The general appearance of this variety is strikingly different from that of the L. nodiflora, in which it is probably included. The L. nodiflora has broad lanceolate leaves, from one to two inches long; the leaves of this variety are spatulate lanceolate and narrow. In the L. nodiflora the peduncles are on erect leafy branches, which might be taken for stems, and they are axillary on their branches; in the new variety, they are also axillary, but rise directly from the creeping stem, which has no branches. The L. nodiflora described in Chapman's Southern Flora is in part of the first variety, and the L. lanceolata of Gray's Manual is that of the new variety I named L. microphylla, which seems to me far preferable, as the nodiflora only has lanceolate leaves, but the microphylla has not.

14. *Asplenium ebeneum*—Ait. var. A.—Bacculum Rubrum.—Stipe and rachis purplish brown, glossy, tall, one to two feet high. Fronds linear, lanceolate, accuminate, pinnate. Pinnæ numerous, sessile auricled on both sides of the base, coarsely serate, the pinnæ below the middle gradually decreasing in length. Fruit-dots elongated, from twenty to thirty on each pinna. Pinnæ distant.

Habitat—Near Baton Rouge, on the margin of cultivated fields, overgrown with trees and bushes. It is also found on the edge of the cane brakes.

PHÆNOGAMOUS OR FLOWERING PLANTS.

DICOTYLEDONOUS PLANTS.

RANUNCULACEÆ—*Crowfoot Family.*

Ranunculus sceleratus, L., Celery Crowfoot, Baton Rouge, East Baton Rouge.

Ranunculus muricatus, L., Prickly-fruited Crowfoot, Baton Rouge, East Baton Rouge.

Ranunculus parviflora, L., Small-flowered Crowfoot, Baton Rouge, East Baton Rouge.

Ranunculus pusillus, Poir, Puny Crowfoot, Baton Rouge, East Baton Rouge.

Clematis crispa, L., Vine, Crisp flowered Clematis, Franklin, St. Mary.

Clematis Virginiana, L., Vine, Virgins' Bower, Baton Rouge, East Baton Rouge.

Clematis Viorna, L., Vine, Leather-flower, Baton Rouge, East Baton Rouge.

Anemone thalictroides var uniflora, Meadow-rue Windflower.

Adonis autumnalis, L., Phaesant's Eye, near Mississippi river, West Baton Rouge.

Delphinium azureum, Mich, Azure Larkspur, Old Seminary, Rapides.

Delphinium Ajacis, Ajax Larkspur, Cultivated.

MAGNOLIACEÆ—*Magnolia Family.*

Magnolia grandiflora, L., tree, Great Laurel Magnolia, Baton Rouge, East Baton Rouge.

Magnolia fuscatus, tree, Sweet scented Magnolia, native of China, Cultivated.

Liriodendron tulipifera, L., tree, Tulip tree White Poplar, Baton Rouge, East Raton Rouge.

Illicium Floridanum Ellis, tree, Florida Anise tree, Port Hudson Railroad, East Feliciana.

NELUMBIACEÆ—*Nelumbo Family.*

Nelumbo luteum Willd, Water Chinquapine (No flower), Prairie Lake, St. Landry.

NYMPHÆACEÆ—*Water Lily Family.*

Nuphar advena Ait, Bonnets, Spatter-Dock, Covington, St. Tammany.

SARRACENIACEÆ—*Pitcher Plant Family.*

Sarracenia rubra Walt, Red Flowered, Trumpet-Leaf, Covington, St. Tammany.

PAPAVARACEÆ—*Poppy Family.*

Agremone Mexicana L, Mexican Poppy, near Mississippi River, West Baton Rouge.

CRUCIFERÆ—*Mustard Family.*

Nasturtium sylvestre R Br., Yellow Cress, New Orleans, Orleans.

Nasturtium officinale R. Br., Water Cress, Baton Rouge, East Baton Rouge.

Nasturtium tanacetifolium, Hook. and Arn. Tansey-Leaved Cress, Baton Rouge, East Baton Rouge.

Nasturtium sessiliflorum, Nutt, Stalkless-flowered Cress, Baton Rouge, East Baton Rouge.

Nasturtium palustre, D. C., Swamp Cress, Baton Rouge, East Baton Rouge.

Cardamine Ludoviciana, Hook. Louisiana Bitter-Cress, Baton Rouge, East Baton Rouge.

Cardamine hirsuta, L., Common Bitter-Cress, Baton Rouge, East Baton Rouge.

Lepidium Virginicum, L., Wild pepper grass, Baton Rouge, East Baton Rouge.

Senebiera pinnatifida, D. C., Wart Cress, Swine Cress, Baton Rouge, East Baton Rouge.

Capsella bursa-pastoris Moench. Shepherd's Purse, Baton Rouge, East Baton Rouge.

Sisymbrium officinale, L., Hedge Mustard, Baton Rouge, East Baton Rouge.

Brassica rapa var. depressa, Turnip, cultivated.

Brassica oleracea, Cabbage, cultivated.

Raphanus sativus, Radish, cultivated.

Cheiranthus cheiri, Wallflower, Southern Europe, cultivated.

CAPPARIDACEÆ—*Caper Family.*

Cleome pungens Willd., Strong-scented Cleome, New Orleans, Orleans.

VIOLACEÆ—*Violet Family.*

Viola cucullata Ait., Hood-leafed Violet, Baton Rouge, East Baton Rouge.

Viola primulæfolia, L., Primrose-leafed Violet, Amite River, East Baton Rouge.

Viola tricolor, Pansey, Heart's ease, cultivated.

HYPERICACEÆ—*St. John's Wort Family.*

Hypericum corymbosum, Muhl., Corymbose St. John's wort, Chicotville, St. Landry.

Hypericum cystifolium, Lam., Rock rose-leafed St. John's wort, Ville Platte, St. Landry.

Hypericum prolificum, L., shrubs, St. John's wort, Old Seminary, Rapides.

Hypericum prolificum, L., var. densiflorum Pursh, Dense-flowered Hypericum, Pontchatoula, Tangipahoa.

Hypericum gladioides, Lam., Sword-leafed Hypericum, Ville Platte, St. Landry.

Hypericum mutilum, L. H. parviflorum, Muhl., small-flowered Hypericum, Old Seminary, Rapides.

Hypericum Canadense, L., Canadian Hypericum, Opelousas, St. Landry.

Hypericum Sarothra, Michx., Orange grass, Pine weed, Old Seminary, Rapides.

Ascyrum stans, Michx., shrubs, St. Peter's wort, Old Seminary, Rapides.

Ascyrum Crux Andrea, L., shrubs, St. Andrew's Cross, Old Seminary, Rapides.

Portulaccaceæ—*Purselane Family.*

Portulacca oleracea, L., Purselane, Baton Rouge, East Baton Rouge.

Caryophyllaceæ—*Pink Family.*

Cerastium valgatum, L., Common Mouse Ear, Baton Rouge, East Baton Rouge.

Stellaria media, Smith, Common Star wort, Baton Rouge, East Baton Rouge.

Arenaria serpyllifolia, L, Thyme-leafed Sandwort, introduced.

Sagina Elliottii, Fenzl, Elliott's Pearlwort, Baton Rouge, East Baton Rouge.

Siphonychia Americana, Tor. and Gray, American Siphonychia, Old Seminary, Rapides.

Silene quinquevulnera, L, Five-wounded Silene (introduced), Amite river, East Baton Rouge.

Silene armeria, L, Sweet William Catchfly (introduced), Baton Rouge, East Baton Rouge.

Mollugo verticillata, L, Carpet weed, Mandeville, St. Tammany.

Dianthus Chinensis, China Pink, Native of China, cultivated.

Malvaceæ—*Mallow Family.*

Hibiscus incanus, Wendl, Hoary Rose Mallow, New Orleans, Orleans.

Hibiscus Moscheutos, L, Swamp Mallow, Brashear City, St. Mary.

Hibiscus militaris, Cav, Halbert-leafed Hibiscus, Brashear City, St. Mary.

Hibiscus coccineus, Walt, Scarlet-flowered Hibiscus, Bayou Portage, St. Mary.

Hibiscus Syriacus, L, Shrubby Althea, Native of Spain, cultivated.

Kosteletzkya Virginica Presl. Hibiscus Virginicus, L, Virginia Hibiscus, Lake Pontchartrain, Orleans.

Callirrhoe papaver, Gray, Poppy-flowered Callirrhoe, Old Seminary, Rapides.

Modiola multifida Moench. Malva Caroliniana, Many cleft-leafed Modiola, Baton Rouge, East Baton Rouge.

Sida spinosa, L, Thorny Sida, Franklin, St. Mary.

Gossypium herbaceum, Cotton, cultivated.

TILLIACEÆ—*Linden Family.*

Corchorus siliquosus, L, Pod-bearing Corchorus, cultivated.

CAMELLIACEÆ—*Camelia Family.*

Camellia Japonica, Japan Rose, Native of Japan, cultivated.

AURANTIACEÆ—*Orange Family.*

Citrus aurantium, tree, Orange, Native of Asia, cultivated.

LINACEÆ—*Flax Family.*

Linum Virginianum, Wild Flax, Old Seminary, Rapides.

OXALIDACEÆ—*Wood Sorrel Family.*

Oxalis stricta, L, Yellow Wood Sorrel, Baton Rouge, East Baton Rouge.

GERANIACEÆ—*Geranium Family.*

Geranium Carolinianum, L, Carolina Cranesbill, Baton Rouge, East Baton Rouge.

Pelargonium inquinans, Scarlet Pelargonium, cultivated.

BALSAMINACEÆ—*Balsam Family.*

Impatiens balsamina, Garden Balsamine, Native of East Indies, cultivated.

RUTACEÆ—*Rue Family.*

Zanthoxylum Carolinianum, Lam, tree, Toothache Tree, Port Hudson and Clinton Railroad, East Feliciana.

Ptelea trifoliata, L, Three-leafed Ptelea, Baton Rouge, East Baton Rouge.

ANACARDIACEÆ—*Cashew Family.*

Rhus copallina, L, shrub, Sumach, Old Seminary, Rapides.

Rhus toxicodendron, L, var. radicaus Tor, vine, Poison Ivy, Baton Rouge, East Baton Rouge.

VITACEÆ—*Vine Family.*

Vitis æstivalis, Michx, Summer Grape, Grosse Tête Railroad, West Baton Rouge.

Vitis vulpina, L, Muscadine, Baton Rouge, East Baton Rouge.

Ampelopsis quinquefolia, Michx. vine, Virginia Creeper, Baton Rouge, East Baton Rouge.

RHAMNACEÆ—*Buckthorn Family.*

Rhamnus lanceolata, Pursh, tree, Lance-leafed Buckthorn, Baton Rouge, East Baton Rouge.

Ceanothus Americanus, L., Jersey tea, Old Seminary, Rapides.

Berchemia volubilis, D. C., vine, Supple Jack, Baton Rouge, East Baton Rouge.

CELASTRACEÆ—*Staff Tree Family.*

Euonymus Americanus, L., shrub, Strawberry bush, Baton Rouge, East Baton Rouge.

SAPINDACEÆ—*Soap-berry Family.*

Cardiospermum Halicacabum, L., Balloon vine Heart-seed, Port Hudson, East Feliciana.

Aesculus Pavia, L., shrub, Small Buckeye, Baton Rouge, East Baton Rouge.

ACERACEÆ—*Maple Family.*

Acer dasycarpum, Ehrh., tree, Silver Maple, Baton Rouge, East Baton Rouge.

POLYGALACEÆ—*Milkwort Family.*

Polygala fastigiata, Nutt, Tower-branched Polygala, Chicotville, St. Landry.

Polygala ramosa, Ell., Branching Polygala, Covington, St. Tammany.

Polygala cruciata, L., Cross-leafed Polygala, Pontchatoula, Tangipahoa.

Polygala Nutallis, Tor. & Gr., Nutall's Polygala, Amite City, Tangipahoa.

Polygala nana, D. C., Dwarf Polygala, Old Seminary, Rapides.

Polygala cymosa, Walt., Bushy Polygala, Pontchatoula, Tangipahoa.

Polygala incarnata, L., Flesh-colored Polygala, Old Seminary, Rapides.

LEGUMINOSÆ—*Bean Family.*

Crotalaria sagittalis, L., Arrow-leafed Rattle Box, Old Seminary, Rapides.

Crotalaria Purshii, D. C., Pursh's Rattle Box, Ponchatoula, Tangipahoa.

Crotalaria ovalis Pursh, Oval-leafed Rattle Box, Covington, St. Tammany.

Melilotus officinalis, L., Sweet Clover, near Mississippi river, West Baton Rouge.

Stylosanthes elatior, Swartz, Tall Pencil-flower, Old Seminary, Rapides.

Vicia sativa, L., Vetch, Baton Rouge, East Baton Rouge.

Trifolium repens, L., White Clover, Baton Rouge, East Baton Rouge.

Trifolium pratense, L., Red Clover, Baton Rouge, East Baton Rouge.

Trifolium Carolinianum, Michx, Carolina Trefoil, Brashear City, St. Mary.

Trifolium procumbens, L., Yellow or Hop Clover, Baton Rouge, East Baton Rouge.

Trifolium agrarium, L., Field Clover.

Desmodium viridiflorum, Bech, Green-flowered Tick-Trefoil, Chicotville, St. Landry.

Desmodium strictum, D. C., Stiff Tick-Trefoil, Chicotville, St. Landry.

Desmodium nudiflorum, D. C., Sparse-flowered Tick-Trefoil, Chicotville, St. Landry.

Desmodium glabellum, D. C., Smooth Tick-Trefoil, Chicotville, St. Landry.

Desmodium molle, D. C., Soft-leafed Tick-Trefoil, Chicotville, St. Landry.

Desmodium tenuifolium, Tor. and Gray, Slender-leafed Tick-Trefoil, Chicotville, St. Landry.

Desmodium canescens, D. C., Hoary Tick-Trefoil, Opelousas, St. Landry.

Desmodium paniculatum, D. C., Cluster-flowered Tick-Trefoil, Baton Rouge, East Baton Rouge.

Lespedeza capitata, Mich., Bush Clover, Old Seminary, Rapides.

Lespedeza hirta, Ell., Rough-stemmed Bush Clover, Old Seminary, Rapides.

Cercis Canadensis, L., tree, Red Bud, Baton Rouge, East Baton Rouge.

Amorpha fruticosa, L., shrub, False Indigo, Baton Rouge, East Baton Rouge.

Phaseolus helvolus, L., Pale-red-flowered Bean, Chicotville, St. Landry.

Phaseolus diversifolius, Pers., Various-leafed Bean, Port Hudson, East Feliciana.

Phaseolus vulgaris, String Bean, Cultivated.

Phaseolus lunatus, Lima Bean, Cultivated.

Vigna glabra, Savi., Smooth Vigna, New Orleans, Orleans.

Baptisia lanceolata, Ell., Lance-leafed Baptisia, Ville Platte, St. Landry.

Centrosema Virginiana, Pursh., Spurred Butterfly Pea, Port Hudson, East Feliciana.

Clitoria Mariana, L., Butterfly Pea, Old Seminary, Rapides.

Tephrosia Virginiana, Pers., Goat's Rue, Old Seminary, Rapides.

Tephrosia spicata, Tor. and Gray, Spike-flowered Tephrosia, Old Seminary, Rapides.

Tephrosia paucifolia, N. sp., Sparse-leafed Tephrosia, Pontchatoula, Tangipahoa.

Tephrosia angustifolia, N. sp., Narrow-leafed Tephrosia, Pontchatoula, Tangipahoa.

Tephrosia multiflora, N. sp., Many-flowered Tephrosia, Pontchatoula, Tangipahoa.

Galactia pilosa, Ell., Hairy Milk pea, Bayou Portage, St. Mary.

Galactia brachypoda, Tor. and Gray, Short-podded Galactia, Covington, St. Tammany.

Galactia mollis, Michx., Soft-leafed Galactia, Pontchatoula, Tangipahoa.

Galactia spiciformis, Tor. and Gray, Spike-flowered Galactia, Grand Isle, Jefferson.

Rhynchosia minima, D. C., Small-flowered Rhynchosia, Baton Rouge, East Baton Rouge.

Rhynchosia tomentosa, Tor. and Gr., var. volubilis, Changeable Woolly Rhynchosia, Old Seminary, Rapides.

Rhynchosia tomentosa, Tor. and Gr., var. erecta, Erect Woolly Rhynchosia, Chicotville, St. Landry.

Sesbania macrocarpa, Muhl., Large-fruited Sesbania, Opelousas, St. Landry.

Robinia pseudacacia, L., tree, Flowering Locust, Baton Rouge, East Baton Rouge.

Gleditchia triacanthos, L., tree, Honey locust, Baton Rouge, East Baton Rouge.

Arachis hypogæa, Peanut, Native of Africa, Cultivated.

Cassia nictitans, L., Wild, Sensitive plant, Chicotville, St. Landry.

Cassia obtusifolia, L., Obtuse-leafed Cassia, Port Hudson, East Feliciana.

Cassia Marylandica, L., Wild Senna, Old Seminary, Rapides.

Cassia chamaecrista, L., Patridge Pea, Ville Platte, St. Landry.

Cassia occidentalis, L., Western Cassia, Baton Rouge, East Baton Rouge.

Schrankia angustata var. brachycarpa, Tor. and Gr., Short-fruited Sensitive Briar, Washington, St. Landry.

Schrankia uncinata, Willd, Prickle-hooked Briar, Old Seminary, Rapides.

Desmanthus brachylobus, D. C., Short-lobed Desmanthus, Port Hudson, East Feliciana.

Albizzia Julibrissin, tree, Tree Mimosa, cultivated.

Acacia Farnesiana, tree, Black Thorn, Baton Rouge, East Baton Rouge.

Acacia dealbata, tree, Phyllode-leafed Acacia, native of Australia, cultivated.

Dolichos lablab, Egyptian Bean, native of Egypt, cultivated.

Wistaria Chinensis, Cl. bush, Chinese Wistaria, native of China, cultivated.

ROSACEÆ—*Rose Family.*

Agrimonia Eupatoria, L., Common Agrimony, Chicotville, St. Landry.

Cratægus spathulata, Michx, tree, Broad-leafed Thorn, Baton Rouge, East Baton Rouge.

Cratægus arborescens, Ell, tree, Tree-like Thorn, Old Seminary, Rapides.

Cratægus aestivalis, Tor. and Gray, tree, Summer Hawthorn, Old Seminary, Rapides.

Cratægus apiifolia, Tor. and Gray, tree, Celery-leafed Hawthorn, Baton Rouge, East Baton Rouge.

Cratægus Crus galli, L., Cockspur Thorn, Baton Rouge, East Baton Rouge.

Potentilla Norwegica, L., Norwegian Cinqfoil, Baton Rouge, introduced.

Rosa laevigata, Michx, bush, Cherokee Rose, Baton Rouge, East Baton Rouge.

Rubus trivialis, Michx, Blackberry, Dewberry, Baton Rouge, East Baton Rouge.

Rubus villosus, Ait., High Blackberry, Baton Rouge, East Baton Rouge.

Prunus Virginianus, L., tree, Wild Cherry tree, Amite River, East Baton Rouge.

Prunus Carolinianus, Ait., tree, Cherry-leafed Wild Orange, Baton Rouge, cultivated.

Prunus domestica, tree, Common Plum, cultivated.

Cydonia Japonica, shrub, Japan Quince, cultivated.

Amygdalus Persica, Mull, tree, Peach tree, cultivated.

Amygdalus nana, shrub, Flowering Almond, native of Russia, cultivated.

Spiræa hypericifolia, shrub, Italian May, St. Peter's Wreath, cultivated.

Calycanthucæ--*Carolina Allspice Family.*

Calycanthus floridus, L., shrub, Flowering Sweet-scented Shrub, native of North Carolina, cultivated.

Melastomaceæ—*Melastoma Family.*

Rhexia Mariana, L., Deer grass, Old Seminary, Rapides.

Rhexia glabella, Michx. Smooth-stemmed Rhexia, Pontchatoula, Tangipahoa.

Rhexia Virginica, L., Virginia Rhexia, Ville Platte, St. Landry.

Rhexia stricta, Pursh, Stiff-stemmed Rhexia, Ville Platte, St. Landry.

MYRTACCEÆ—*Myrtle Family.*

Punica granatum, tree, Pomegranate, native of South Europe, naturalized.

LYTHRACEÆ—*Loose-strife Family.*

Ammannia ramasior, Michx, Many-branched Ammania, Baton Rouge, East Baton Rouge.

Ammannia humilis, Michx, Low-stemmed Ammania, Clinton, East Feliciana.

Ammannia latifolia, L., Wide-leafed Ammania, Washington, St. Landry.

Lythrum alatum, Pursh, Loose-strife, Franklin, St. Mary.

Cuphea vicossissima, Jacq., Clammy Cuphea.

Lagerstræmia Indica, L., Grape Myrtle, Ville Platte, naturalized.

ONAGRACEÆ—*Evening Primrose Family.*

Œnothera biennis, L. O. grandiflora Ait., Common Evening Primrose, Washington, St. Landry.

Œnothera fruticosa, L., Sun-drops Primrose, Amite City, Tangipahoa.

Œnothera sinuata, L., Sinuate-leafed Œnothera, Baton Rouge, East Baton Rouge.

Œnothera sinuata, var. humifusum, Tor. & G., Lowly Œnothera, Grand Isle, Jefferson.

Œnothera paludosa, N. Sp., Swamp Œnothera, Pine prairie swamp, St. Landry.

Ludwigia alternifolia, L., Seed Box, Chicotville, St. Landry.

Ludwigia hirtella, Raf., Rough-haired Ludwigia, Pontchatoula, Tangipahoa.

Ludwigia palustris, Ell., Water Purselane, New Orleans, Orleans.

Jussiæa decurrens, D. C., Stem-winged Jussiæa, Baton Rouge, East Baton Rouge.

Jussiæa leptocarpa, Nutt, Slender-fruited Jussiæa, Port Hudson East Feliciana.

Jussiæa grandiflora, Michx, Large-flowered Jussiæa, Grosse Tete Railroad, West Baton Rouge.

Jussiæa Boydiana, N. Sp., Boyd's Jussiæa, Port Hudson, East Feliciana.

Gaura angustifolia, Michx, Narrow-leafed Gaura, Old Seminary, Rapides.

Gaura biennis, L., Biennial Gaura, Old Seminary, Rapides.

Gaura filipes, Spach., Slender-stalked Gaura, Covington, St. Tammany.

CACTACEÆ—*Cactus Family.*

Opuntia Ficus Indica, Haw., Indian Fig Prickly Pear, Franklin, St. Mary.

Opuntia vulgaris, Mill., Common Prickly Pear, Amite river, East Baton Rouge.

PASSIFLORACEÆ—*Passion Flower Family.*

Passiflora lutea, L., Yellow Passion Flower, Port Hudson, East Feliciana.

Passiflora incarnata, L., Passion Flower, Maypop, near Mississippi river, West Baton Rouge.

CUCURBITACEÆ—*Cucumber Family.*

Melothria pendula, L., Hanging Melothria, Opelousas, St. Landry.

Lagenaria vulgaris, Ser., Gourd, native of Tropics, cultivated.

Momordica balsamina, Common Balsam Apple, native of East Indies, cultivated.

Cucurbita pepo, Pumpkin, native of the Levant, cultivated.

Cucurbita verucosa, Warted Squash, cultivated.

Cucumis citrulus, Ser., Water Melon, native of Africa and India, cultivated.

CRASSULACEÆ—*Orpine Family.*

Penthorum sedoides, L., Ditch Stone Crop, Grosse Tête, Baton Rouge, West Baton Rouge.

SAXIFRAGACEÆ—*Saxifrage Family.*

Itea Virginica, L., tree, Virginia Itea, Baton Rouge, cultivated.

Hydrangea quercifolia Bartram, Oak-leafed Hydrangea, Baton Rouge, East Baton Rouge.

Hydrangea hortensia, Garden Hydrangea, cultivated.

Deutzia crenata, shrub, Notch-leafed Deutzia, cultivated.

Philadelphus coronarius, shrub, Mock Orange, False Syringa, native of South Europe, cultivated.

HAMAMELACEÆ—*Witch Hazel Family.*

Liquidamber Styraciflua, L., tree, Sweet Gum, Old Seminary, Rapides.

UMBELLIFERÆ—*Parsley Family.*

Leptocaulis divaricatus, D. C., Spread-leafed Leptocaulis, Baton Rouge, East Baton Rouge.

Daucus pusillus, Mich., Slender Carrot, Baton Rouge, East Baton Rouge.

Archemora rigida, D. C., Stiff-leafed Cowbane, Old Seminary, Rapides.

Sanicula Canadensis, L., Canada Black Snakeroot, Old Seminary, Rapides.

Discopleura capillacea, D. C., Fine-leafed Mock-bishop weed, Old Seminary, Rapides.

Eryngium Baldwinii Spreng, Baldwin's Eryngium, Pontchatoula, Tangipahoa.

Chærophyllum Teinturieri, Hook and Arn, Teinturier's Chervil, Atchafalaya river, St. Mary.

Chærophyllum procumbens, Lam, Reclining Chervil, Grosse Tete, West Baton Rouge.

Hydrocotyle interrupta Muhl. Common Pennywort, Washington, St. Landry.

Hydrocotyle umbellata L., Umbellate Pennywort, Clinton, East Feliciana.

Apium graveolens, Celery, Native of Britain, cultivated.

Fœniculum vulgare, Common Fennel, Native of England.

Conselinum Canadense, Tor. and Gr., Canada Hemlock Parsley, Brashear City, St. Mary.

Cicuta maculata, L, Water Hemlock, Old Seminary, Rapides.

CORNACEÆ—*Dogwood Family.*

Cornus florida, L, tree, Flowering Dogwood, Baton Rouge, East Baton Rouge.

Cornus stricta, Lam, tree, Stiff Cornel, Grosse Tete Railroad, West Baton Rouge.

CAPRIFOLIACEÆ—*Honeysuckle Family.*

Sambucus Canadensis, L, shrub, Canadian Elder, Baton Rouge, East Baton Rouge.

Lonicera sempervirens, Ait, vine, Evergreen Woodbine, Baton Rouge, East Baton Rouge.

Viburnum prunifolium, L, shrub, Black Haw, Baton Rouge, East Baton Rouge.

RUBIACEÆ—*Madder Family.*

Mitchella repens, L, Creeping Partridge berry, Amite river, East Baton Rouge.

Mitreola sissilifolia, Tor. and Gr., Stalkless-leafed Miterwort, Pontchatoula, Tangipahoa.

Mitreola petiolata, Tor. and Gr., Stalk-leafed Miterwort, Ville Platte, St. Landry.

Polypremum procumbens, L, Reclining Polypremum, Old Seminary, Rapides.

Spermacoce glabra, Michx, Smooth Buttonweed, Washinton, St. Landry.

Cephalanthus occidentalis, L, Button Bush, Old Seminary, Rapides.

Diodia teres Walt Spermacoce diodina, Michx, Round-stemmed Buttonweed, Old Seminary, Rapides.

Diodia Virginiana, L, Virginia Diodia, Baton Rouge, East Baton Rouge.

Galium circaesans, Michx, Wild Liquorice, Washington, St. Landry.

Galium aparine, L, Goose-grass, Cleavers, near Mississippi river, West Baton Rouge.

Oldenlandia cœrulea, Gray, Houstonia cœrulea, Bluets, Amite river, East Baton Rouge.

Oldenlandia angustifolia, Gray, Narrow-leafed Oldenlandia, Old Seminary, Rapides.

Oldenlandia angustifolia var. filifolia, Gr., Thread-leafed Oldenlandia, Old Seminary, Rapides.

Gelsemium sempervirens, Ait, Yellow Jessamine, Baton Rouge, East Baton Rouge.

Spigelia Marylandica, L, Pinkroot, Cheneyville, Rapides.

COMPOSITÆ—*Composite Family.*

Vernonia fasciculata, Michx, Bushy-branched Ironweed, Opelousas, St. Landry.

Vernonia angustifolia, Michx, Narrow-leafed Ironweed, Chicotville, St. Landry.

Elephanthopus Carolinianus, Willd, Carolina Elephant's Foot, Old Seminary, Rapides.

Carphephorus bellidifolia, Tor. and Gr., Daisy-leafed Carphephorus, Amite City, Tangipahoa.

Liatris gracilis, Pursh, Slender-stemmed Liatris, Amite City, Tangipahoa.

Liatris squarrosus, Willd, Blazing Star, Pontchatoula, Tangipahoa.

Liatris elegans, Willd, Elegant-flowered Liatris, Old Seminary, Rapides.

Pluchea bifrons, D. C., Double-leafed Marsh Fleabane, Old Seminary, Rapides.

Pluchea fœtida, D. D., Strong-scented Fleabane, Old Seminary, Rapides.

Baccharis hamilifolia, L, shrub, Sea Groundsel Tree, Baton Rouge, East Baton Rouge.

Ambrosia trifida, L, Great Ragweed, Baton Rouge, East Baton Rouge.

Ambrosia artemissiæfolia, L, Roman Wormwood, Hogweed, Baton Rouge, East Baton Rouge.

Lappa major, Gaert, Burdock, Baton Rouge, East Baton Rouge.

Iva frutescens, L, shrub, Shrubby Marsh Elder, Bayou Portage, St. Mary.

Pterocaulon pychnostachyum, Ell., Close-flowered Pterocaulon.

Parthenium hysterophorus, L., Late-blooming Parthenium, New Orleans, Orleans.

Gnaphalium purpureum, L., Purplish Cudweed, Baton Rouge, East Baton Rouge.

Gnaphalium polycephalum, Michx., Common Everlasting, Old Seminary, Rapides.

Cirsium altissimum, Spreng., High-stemmed Plumed Thistle, Baton Rouge, East Baton Rouge.

Cirsium muticum, Michx., Swamp Thistle, Old Seminary, Rapides.

Cirsium Virginianum, Michx., Virginia Thistle, Baton Rouge, East Baton Rouge.

Conoclinium coelestinum, D. C., Blue Mistflower, Old Seminary, Rapides.

Eupatorium teucrifolium, Willd., Germander-leafed Eupatorium, Chicotville, St. Landry.

Eupatorium album, L., White-flowered Eupatorium, Old Seminary, Rapides.

Eupatorium perfoliatum, L., Boneset Thoroughwort, Old Seminary, Rapides.

Eupatorium serotinum, Michx., Late-blooming Eupatorium, Baton Rouge, East Baton Rouge.

Eupatorium ivaefolia, L., Ivy-leafed Eupatorium, Baton Rouge, East Baton Rouge.

Eupatorium incarnatum, Walt., Rose-flowered Eupatorium, Baton Rouge, East Baton Rouge.

Eupatorium coronopifolium, Willd., Buckshorn-leafed Eupatorium, Baton Rouge, East Baton Rouge.

Eupatorium fœniculaceum, Willd. (Introduced since the war), Fennel-leafed Eupatorium, Baton Rouge, East Baton Rouge.

Eupatorium rotundifolium, L., Round-leaved Eupatorium, Chicotville, St. Landry.

Eupatorium parviflorum, Ell., Small-flowered Eupatorium, Ville Platts, St. Landry.

Eupatorium ageratoides, L., Nettle-leafed Eupatorium, Old Seminary, Rapides.

Eupatorium hyssopifolium, L., Narrow-leafed Eupatorium.

Mikania scandens, Willd., Climbing Hempweed, Old Seminary, Rapides.

Kuhnia eupatoroides, L., Eupatorium-flowered Kuhnia, Old Seminary, Rapides.

Stokesia cyanea, L'Her., Violet-flowered Stokesia, Covington, St. Tammany.

Achillea millefolia, L., Common Yarrow, Baton Rouge, East Baton Rouge.

Seriocarpus conyzoides, Nees., White-topped Aster.

Seriocarpus tortifolius, Nees., Wave-leafed Seriocarpus.

Diplopappus umbellatus, Tor. and Gr., Double-bristled Aster, Old Seminary, Rapides.

Diplopappus amygdalinus, Tor. and Gr., Almond-leafed Bristled Aster.

Eclipta longifolia, Schrad., Long-leafed Eclipta, New Orleans, Orleans.

Eclipta erecta, L., Erect Eclipta, Baton Rouge, East Baton Rouge.

Maruta Cotula, D. C., Mayweed, Baton Rouge, East Baton Rouge.

Erigeron Canadense, L., Canada Fleabane Butterweed, Old Seminary, Rapides.

Erigeron Philadelphicum, L., Fleabane, Baton Rouge, East Baton Rouge.

Erigeron strigosum, Muhl., White-weed, Daisy, Ville Platte, St. Landry.

Erigeron bellidifolium, Muhl., Robin's Plantain, Grand Isle, Jefferson.

Chrysanthemum parthenium, Feverfew, cultivated.

Chrysanthemum coronarium, Garden Chrysanthemum, cultivated.

Aster dumosus, L., Bushy Aster, Baton Rouge, East Baton Rouge.

Aster longifolius, Lam, Long-leafed Aster, Covington, St. Tammany.

Aster miser, L., Starved Aster, Old Seminary, Rapides.

Aster Baldwinii, Tor. and Gray, Baldwin's Aster, Old Seminary, Rapides.

Aster multiflorus, Ait, Many-flowered Aster, Old Seminary, Rapides.

Aster ericoides, L., var. villosus, Woolly-stemmed Aster, Old Seminary, Rapides.

Aster sericeus, Vent, Silk-leafed Aster, Old Seminary, Rapides.

Aster concolor, L., One-colored Aster, Old Seminary, Rapides.

Aster patens, L., Spreading Aster, Old Seminary, Rapides.

Aster adnatus, Nutt, Adherent-leafed Aster, Old Seminary, Rapides.

Aster squarrosus, Walt, Ragged Aster, Old Seminary, Rapides.

Aster lævis, L., Smooth Aster.

Aster simplex, Willd., Willow-leafed Aster.

Aster Tradescanti, L., Tradescant's Aster.

Aster diversifolius, Michx, Varied-leafed Aster.

Aster surculosus, Michx, Sucker Aster.

Aster ericoides, L., Heath Aster.

Solidago arguta, Ait, Sharp-notched Golden Rod, Baton Rouge, East Baton Rouge.

Solidago flavovirens, Chapm., Yellow and Green-flowered Solidago, Old Seminary, Rapides.

Solidago virgata, Ell., Straight-stemmed Solidago, Old Seminary, Rapides.

Solidago odora, Ait, Sweet-scented Golden Rod, Old Seminary, Rapides.

Solidago altissima, L., Tall Golden Rod, Old Seminary, Rapides.

Solidago rugosa, Willd, Rough-leafed Golden Rod, Old Seminary, Rapides.

Solidago gigantea, Ait, Gigantic Golden Rod, Port Hudson, East Feliciana.

Gaillardia lanceolata, Michx., Lance-leafed Gaillardia, Old Seminary, Rapides.

Borrichia frutescens, D. C., Sea Ox-Eye, Grand Isle, Jefferson.

Helenium quadridentatum, Lab., Four-toothed Helenium, Baton Rouge, East Baton Rouge.

Helenium Seminariense, N. Sp., Seminary Helenium, Old Seminary, Rapides.

Helenium tenuifolium, Nutt., Slender-leaved Bitter-weed, Old Seminary, Rapides.

Chrysopsis trichophylla, Nutt., Hairy-leaved Golden Aster, Old Seminary, Rapides.

Chrysopsis argentea, Nutt., Silver-leafed Golden Aster, Old Seminary, Rapides.

Chrysopsis graminifolia, Nutt., Grass-leafed Golden Aster, Chicotville, St. Landry.

Spilanthes Nuttallii, Tor. & Gray, Nuttall's Spilanthes, Baton Rouge, East Baton Rouge.

Spilanthes repens, Mich., Creeping Spilanthes, Old Seminary, Rapides.

Baldwinia uniflora, Ell., One-flowered Baldwinia, Ville Platte, St. Landry.

Actinomeris helianthoides, Nutt., Sunflower-like Actinomeris, Old Seminary, Rapides.

Bidens frondosa, L., Leafy Beggar's Tick, Old Seminary, Rapides.

Bidens chrysanthemoides, Mich., Chrysanthemum, Old Beggar's Tick, Old Seminary, Rapides.

Coreopsis lanceolata, L., Lance-leafed, Coreopsis, Old Seminary, Rapides.

Coreopsis trichosperma, Mich., Tickseed Sunflower, Old Seminary, Rapides.

Coreopsis tripteris, L., Tall Coreposis, Chicotville, St. Landry.

Coreopsis tinctoria, L., Dyer's Coreopsis, cultivated.

Coreopsis auriculata, L., Ear-lobed Coreopsis, Chicotville, St. Landry.

Rudbeckia triloba, L., Three-lobed Coneflower, Covington, St. Tammany.

Rudbeckia hirta, L., Rough-leafed Coneflower, Old Seminary, Rapides.

Rudbeckia maxima, Nutt., Tall Coneflower, New Orleans, Orleans.

Helianthus devaricatus, L.. Spreading Sunflower, Old Seminary, Rapides.

Helianthus hirsutus, Raf., Hairy-leafed Helianthus, Pontchatoula, Tangipahoa.

Helianthus rigidus, Desf., Stiff-leafed Helianthus, Old Seminary, Rapides.

Helianthus angustifolis, L., Narrow-leafed Helianthus, Old Seminary, Rapides.

Helianthus annuus, L., Common Sunflower, native of South America; cultivated.

Helianthus tuberosus, L., Jerusalem Artichoke, native of Brazil; cultivated.

Zinnia elegans var. violacea, Elegant Zinnia, native of Mexico, Cultivated.

Zinnia paucifloræ, Few-flowered Zinnia, cultivated.

Silphium asteriscus, L., Starry Silphium, Old Seminary, Rapides.

Polymnia udedalia, L., Yellow Leaf-Cup, Baton Rouge, East Baton Rouge.

Verbesina Virginica, L., Virginia Crown Beard, Baton Rouge, East Baton Rouge.

Senecio lobatus, Pers., Lobe-leafed Senecio, Baton Rouge, East Baton Rouge.

Soliva nasturtiifolia, D. C., Cress-leafed Soliva, Baton Rouge, East Baton Rouge.

Hieracium venosum, L., Rattlesnake-weed, Clinton, East Filiciana.

Hieracium Gronovii, L., Hairy Hawkweed, Baton Rouge, East Baton Rouge.

Krigia Virginica Wild, Virginia Dwarf Dandelion, Baton Rouge, East Baton Rouge.

Pyrrhopappus Carolinianus, D. C., Carolina False Dandelion, Baton Rouge, East Baton Rouge.

Mulgedium Floridanum, D. C., Florida Blue Lettuce, Baton Rouge, East Baton Rouge.

Mulgedium accuminatum, D. C., Point-leafed Blue Lettuce, Baton Rouge, East Baton Rouge.

Erechthites hieracifolia, Raf., Fireweed, Old Seminary, Rapides.

Sonchus asper, Willd, Spiny-leafed Sow Thistle, Baton Rouge, East Baton Rouge.

Lactuca elongata, Muhl, Wild Lettuce, Port Hudson, East Feliciana.

Lactuca sativa, L., Garden Lettuce, Cultivated.

Dahlia variabilis, Desf., Dahlia, native of Mexico, Cultivated.

Centaurea Cyanus, Blue bottle, native of Europe, Cultivated.

Lobeliaceæ—*Lobelia Family.*

Lobelia cardinalis, L., Cardinal Flower, Old Seminary, Rapides.

Lobelia glandulosa, Walt., Glandular Lobelia, Amite City, Tangipahoa.

Lobelia paludosa, Nutt., Swamp Lobelia, Amite City, Tangipahoa.

Lobelia Nuttallii, R. & S., Nuttall's Lobelia, Old Seminary, Rapides.

Lobelia spicata, Lam., Spike-flowered Lobelia, Old Seminary, Rapides.

Lobelia Claytonia, Michx., Clayton's Lobelia, Old Seminary, Rapides.

Lobelia amæna, Michx., Fair-flowered Lobelia, Old Seminary, Rapides.

Campanulaceæ—*Campanula Family.*

Specularia perfoliata, D. C., Campanula perfoliata, L., Venus Looking Glass, Baton Rouge, East Baton Rouge.

Ericaceæ—*Heath Family.*

Rhododendron multiflorum, Tor., Azalea nudiflora, shrub, Purple Azalea, Amite River, East Baton Rouge.

Vaccinium tenellum, Ait., shrub, Tender-branched Blueberry, Covington, St. Tammany.

Oxydendron arboreum, D. C., Andromeda arborea, L., tree, Tree Sorrel, Old Seminary, Rapides.

Aquifoliaceæ—*Holly Family.*

Ilex opaca, Ait., tree, American Holly, Baton Rouge, East Baton Rouge.

Ilex Cassine, L., shrub, Yaupon, Amite River, East Baton Rouge.

STYRACEÆ—*Storax Family.*

Halesia tetraptera, L., shrub, Snowdrop tree, Amite River, East Baton Rouge.

PLANTAGINACEÆ—*Plantain Family.*

Plantago major, L., Common Plantain, Baton Rouge, East Baton Rouge.

Plantago Virginica, L., Virginia Plantain, Baton Rouge, East Baton Rouge.

PLUMBAGINACEÆ—*Leadwort Family.*

Statice lemonium, L., Marsh Rosemary, Grand Isle, Jefferson.

PRIMULACEÆ—*Primrose Family.*

Samolus floribundus, Kunth, Many-flowered Samolus, Baton Rouge, East Baton Rouge.

Lysimachia ciliata, L., Fringed Loose-strife, Old Seminary, Rapides.

LENTIBULACEÆ—*Bladderwort Family.*

Urticularia fibrosa, Walt., Fibre-leafed Bladderwort, Amite City, Tangipahoa.

BIGNONIACEÆ— *Bignonia Family.*

Bignonia capreolata, L., vine, Cross Vine, Baton Rouge, East Baton Rouge.

Tecoma radicans Jus. Bignonia radicans, vine, Trumpet Flower, near Mississippi river, West Baton Rouge.

Martynia proboscidea, Glox., Unicorn Plant, Amite City, Tangipahoa.

Catalpa bignonoides, tree, Catalpa, Opelousas, St. Landry.

SCROPHULARIACEÆ—*Figwort Family.*

Scrophularia nodosa, L., Common Figwort, Baton Rouge, East Baton Rouge.

Gratiola Virginiana, L., Virginia Hedge Hyssop, Clinton, East Feliciana.

Gratiola Floridana, Nutt, Florida Gratiola, Amite City, Tangipahoa.

Herpestis Monnieria, Kunth, Monierias' Herpestis, New Orleans, Orleans.

Herpestis amplexicaulis, Pursh, Clasped-leaved Herpestis, Covington, St. Tammany.

Micranthemum orbiculatum, Michx., Round-leafed Micranthemum, New Orleans, Orleans.

Buchnera Americana, L., Blue Hearts, Old Seminary, Rapides.

Buchnera elongata, Swartz, Long-leafed Buchnera, Pontchatoula, Tangipahoa.

Verbascum Thapsus, L., Mullein, Ville Platte, St. Landry.

Dasystoma quercifolia Benth Gerardia quercifolia, L., Smooth False Foxglove, Chicotville, St. Landry.

Mimulus alatus, Ait., Stem-winged Monkey flower, Ville Platte, St. Landry.

Penstemon digitalis, var. multiflorus, Benth., Many-flowered Penstemon, Old Seminary, Rapides.

Gerardia filifolia, Nutt., Narrow-leafed Gerardia, Old Seminary, Rapides.

Gerardia maritima, Raf., Sea-side Gerardia, Grand Isle, Jefferson.

Gerardia purpurea, L., Purple Gerardia, Old Seminary, Rapides.

Linaria Canadense, Spreng, Wild Toad-Flax, Baton Rouge, East Baton Rouge.

Veronica peregrina, L., Neckweed, Baton Rouge, East Baton Rouge.

Veronica agrestis, Field Speedwell, Baton Rouge, East Baton Rouge.

ACANTHACEÆ—*Acanthus Family.*

Diptheracanthus noctiflorus, Nees., Nightflowering Ruellia, Ville Platte, St. Landry.

Diptheracanthus strepens, Nees., Ruellia strepens, Rough-leafed Ruellia, Baton Rouge, East Baton Rouge.

Dianthera ovata, Walt., Ovate-leafed Water willow, Washington, St. Landry.

Dianthera ovata var. lanceolata, Walt., Lance-leafed Water willow, Baton Rouge, East Baton Rouge.

VERBENACEÆ—*Vervain Family.*

Lippia nodiflora, Michx., Fog Fruit, Baton Rouge, East Baton Rouge.

Lippia nodiflora, var. microphylla, N. Var. Small-leafed Fog fruit Brashear City, St. Mary.

Calycarpa Americana, L., shrub, French Mulberry, Brashear City, St. Mary.

Verbena Aubletia, L., Aubletia's Verbena, Franklin, St. Mary.

Verbena stricta, Nutt., Mullein-leafed Verbena, near Mississippi river, West Baton Rouge.

Verbena urticifolia, L., Nettle-leafed Verbena, Baton Rouge, East Baton Rouge.

Verbena officinalis, L., Vervain, Baton Rouge, East Baton Rouge.

Verbena chamædrifolia, Scarlet Verbena, Cultivated.

Verbena phlogiflora, Crimson Verbena, Cultivated.

Verbena incisa, Cut-leafed Verbena, Cultivated.

Lantana odorata, Sweet scented Lantana, Cultivated.

Lantana camara, Vaulted-flowered Lantana, Native of South America, Cultivated.

LABIATÆ—*Mint Family.*

Hyptis radiata, Willd, Ray-flowered Hyptis, Opelousas, St. Landry.

Mentha piperita, L., Peppermint, Clinton, East Feliciana.

Brunella vulgaris, L., Common Self-Heal, Old Seminary, Rapides.

Nepeta glechoma, Benth, Ground Ivy, Baton Rouge, East Baton Rouge.

Monarda punctata, L., Spotted Horse mint, Old Seminary, Rapides.

Monarda didyma, L., Oswego Tea, Old Seminary, Rapides.

Pyncnanthemum muticum, Pers., Beardless Mountain Mint, Old Seminary, Rapides.

Pycnanthemum linifolium, Pursh, Flax-leafed Mountain Mint, Old Seminary, Rapides.

Salvia azurea, L., Azure-flowered Sage, Old Seminary, Rapides.

Salvia lyrata, L., Lyre-leafed Sage, Baton Rouge, East Baton Rouge.

Salvia fulgens, Bright-colored Sage, native of Mexico, cultivated.

Salvia splendens, Crimson-flowered Sage, native of Mexico, cultivated.

Calamintha nepeta, Link, Basil Thyme, near the Mississippi river, West Baton Rouge.

Leonotis nepetæfolia, R. Br., Catnip-leafed Leonotis, Opelousas, St. Landry.

Scutellarea integrifolia, L., Entire-leafed Scullcap, Clinton, East Feliciana.

Physostegia Virginiana, Benth, False Dragon Head, Amite City, Tangipahoa.

Stachys aspera, Michx, Rough Hedge Nettle, New Orleans, Orleans.

Teucrium Canadense, L., American Germander, Baton Rouge, East Baton Rouge.

Rosemarinus officinalis, Rosemary, cultivated.

Lycopus Virginicus, L., Water Hoarhound, Cheneyville, Rapides.

BORRAGINACEÆ—*Borrage Family.*

Heliotropum Curassavicum, L., Sea-coast Heliotrope, Lake Pontchartrain, Orleans.

Heliophytum Indicum, D. C., India Heliotrope, Baton Rouge, East Baton Rouge.

Myositis verna. Nutt var. macrospermum, Large-seeded Mouse Edr., Baton Rouge, East Baton Rouge.

HYDROPHYLLACEÆ—*Water-leaf Family.*

Nemophila microcalyx, Tesh & Meyer, Small-calyxed Nemophila, Baton Rouge, East Baton Rouge.

HYDROLEACEÆ—*Hydrolea Family.*

Hydrolea quadrivalvus, Walt, Four-valved Hydrolea, Amite City, Tangipahoa.

Hydrolea leptocaulis, N. Sp., Smooth-stemmed Hydrolea, Washington, St. Landry.

Hydrolea Ludoviciana, N. Sp., Louisiana Hydrolea, Bayou Portage, St. Mary.

Nama Jamaicensis, L., Jamaica Nama, Baton Rouge, East Baton Rouge.

POLEMONIACEÆ—*Polemonium Family.*

Phlox Drummondii, Drummond's Phlox—Cultivated.

Phlox glaberrima, Ohio Phlox—Cultivated.

Pyxidanthera barbulata, Michx, Bearded Pyxidanthera.

CONVOLVULACEÆ—*Convulvulus Family.*

Phrabitis nil, Chois., Morning Glory, Baton Rouge, East Baton Rouge.

Phrabitis hispida, Chois., Hairy Morning Glory, Baton Rouge, East Baton Rouge.

Ipomœa tamnifolia, L., Cut-leafed Ipomœa, Ville Platte, St. Landry.

Ipomœa lacunosa, L., Ditch Ipomœa, Baton Rouge, East Baton Rouge.

Ipomœa commutata, Roem. and Shr., Altered Ipomœa, Baton Rouge, East Baton Rouge.

Ipomœa pes-capra, Sweet, Goat-foot-leafed Ipomœa, Grand Isle, Jefferson.

Ipomœa sagittifolia, Bot. Reg., Arrow-leafed Ipomœa, Mandeville, St. Tammany.

Ipomœa pandurata, Meyer, Wild Potato Vine, Old Seminary, Rapides.

Batatas littoralis, Chois., Seacoast Wild Potato, Grand Isle, Jefferson.

Cuscuta Gronovii, Willd., American Dodder, Washington, St. Landry.

Dichondra repens, Forst., var. Caroliniensis, Chois., Carolina Dichondra, Baton Rouge, East Baton Rouge.

SOLANACEÆ—*Nightshade Family.*

Solanum nigrum, L., Common Nightshade, Baton Rouge, East Baton Rouge.

Solanum Caroliniense, L., Horse nettle, Baton Rouge, East Baton Rouge.

Solanum aculeatissimum. Jaq., Prickly leafed Nightshade, Carrollton, Orleans.

Solanum Pseudo capsicum, Jerusalem Cherry, Comite river, East Feliciana.

Physalis viscosa, L., Clammy-leafed Ground Cherry, Port Hudson, East Feliciana.

Physalis pubescens, L., Hairy-leafed Ground Cherry, New Orleans, Orleans.

Physalis lanceolata, Michx., Lance-leafed Ground Cherry, Clinton, East Feliciana.

Datura Strammonium, L., Thornapple, Baton Rouge, East Baton Rouge.

Nicotiana tabacum, L., Tobacco, Cultivated.

Nicotiana longiflora (?), Long-leafed Nicotiana, Brashear City, Introduced.

Lycopersicum esculentum, Mill., Tomato, Cultivated.

Petunia violacea, Violet Petunia, Native of Brazil, Cultivated.

Capsicum annuum, L., Red Pepper, Native of India, Cultivated.

GENTIANACEÆ—*Gentian Family.*

Gentiana ochroleuca, Frœl, Straw-colored Gentian, Old Seminary, Rapides.

Sabbatia stellaris, Pursh., Star-flowered Sabbatia, Grand Isle, Jefferson.

Sabbatia gracilis, Pursh., Slender Sabbatia, Pontchatoula, Tangipahoa.

Sabbatia angularis, Pursh, American Centaury, Chicotville, St. Landry.

Sabbatia calycosa, Pursh., Cup-flowered Sabbatia, Grand Isle, Jefferson.

Sabbatia nana, N., species, Dwarf Sabbatia, Grand Isle, Jefferson.

Sabbatia oligophylla, N., species, Sparse-leafed Sabbatia, Chicotville, St. Landry.

Eustoma exaltatum, Grisebe, High-stemmed Eustoma, Grand Isle, Jefferson.

APOCYNACEÆ—*Dogbane Family.*

Fosteronia difformis, D. C., vine, Unsightly Fosteronia, Old Seminary, Rapides.

Vinca major, Great Periwinkle, Native of Europe, Cultivated.

Nerium Oleander, shrub, Rose Bay Tree, Native of Palestine, Cultivated.

ASCLEPIADACEÆ—*Milkweed Family.*

Asclepias tuberosa, L., Butterfly-weed, Old Seminary, Rapides.

Asclepias verticilata, L., Whorl-leafed Asclepias, Pontchatoula, Tangipahoa.

Asclepias perennis, Willd., Perennial Asclepias, Franklin, St. Mary.

Asclepias paupercula, Michx., Poor Asclepias, Covington, St. Tammany.

Asclepias obtusifolia, Michx., Obtuse-leafed Asclepias, Old Seminary, Rapides.

Seutera maritima Decais. Lyonis maritima, Ell., Seaside Seutera, Grand Isle, Jefferson.

Oleaceæ—*Olive Family.*

Olea fragrans, tree, Fragrant Olive tree, native of China, cultivated.

Syringa vulgaris, tree, Common Lilac, native of Hungary, cultivated.

Forsythia viridissima, Green-leafed Forsythia, native of China, cultivated.

Aristolochaceæ—*Birthwort Family.*

Aristolochia serpentaria, L. A. hastata Nutt., Snake Birthwort, Amite river, East Baton Rouge.

Nyctaginaceæ—*Four O'Clock Family.*

Mirabilis Jalapa, Common Mirabilis,native of West Indies; cultivated.

Phytolaccaceæ—*Pokeweed Family.*

Phytolacca decandra, L., Pokeweed, Baton Rouge, East Baton Rouge.

Chenopodiaceæ—*Goosefoot Family.*

Chenopodium album, L., Pigweed, Franklin, St. Mary.

Chenopodium anthelminticum, L., Worm seed, Baton Rouge, East Baton Rouge.

Sueda maritima, Moq., Seaside Chenopodium, Grand Isle, Jefferson.

Salicornia ambigua, Michx., Samphire, Grand Isle, Jefferson.

Amarantaceæ—*Amaranth Family.*

Amarantus spinosus, L., Thorny Amaranth, Old Seminary, Rapides.

Amarantus hybridus, L., Hybrid Amaranth, Baton Rouge, East Baton Rouge.

Amarantus chlorostachys, Willd., Green-flowered Amaranth, Grand Isle, Jefferson.

Alternanthera achyrantha, R. Br., Chaff-flowered Alternanthera, New Orleans, Orleans.

Iresine celosoides, Ell. I. diffusa H. and B., Cockscomb Iresine, Baton Rouge, East Baton Rouge.

Gomphrena globosa, L., Globe Amaranth, native of India cultivated.

Celosia, cristata, Cockscomb, native of Japan, cultivated.

Amarantus paniculatus, Panicled Amaranth, native of South America, cultivated.

POLYGONACEÆ—*Buckwheat Family.*

Brunnichia cirrhosa, Banks, vine, Tendril-branched Brunnichia, Baton Rouge, East Baton Rouge.

Rumex pulcher, L., Fair-leafed Dock, Baton Rouge, East Baton Rouge.

Rumex crispus, L., Crip-leafed Dock, Baton Rouge, East Baton Rouge.

Polygonum aviculare, L., Goose grass, Knot grass, Baton Rouge, East Baton Rouge.

Polygonum incarnatum, Ell., Rose-flowered Polygonum, Brashear City, St. Mary.

Polygonum hydropiperoides, Michx., Wild Water Pepper, Baton Rouge, East Baton Rouge.

Polygonum amphibium, L., var. aqudticum, Water Persicaria, Baton Rouge, East Baton Rouge.

Polygonum densiflorum, Meis., Close-flowered Polygonum, Washington, St. Landry.

Polygonum setaceum, Baldw., Bristle-sheathed Polygonum, Baton Rouge, East Baton Rouge.

Polygonum Virginianum, L., Virginia Polygonum, Washington, St. Landry.

Polygonum acre, Kunth, Wild Smartweed, Old Seminary, Rapides.

Polygonum Pennsylvanicum, L., Pennsylvania Knotgrass, Old Seminary, Rapides.

Polygonum dumotorum, L., Climbing Polygonum, Cheneyville, Rapides.

LAURACEÆ—*Laurel Family.*

Sassafras officinale, Nees, Laurus Sassafras, L. tree, Sassafras, Baton Rouge, East Baton Rouge.

LORANTHACEÆ—*Mistletoe Family.*

Phoradendron flavescens, Nutt. Viscum flavescens, shrub, Mistletoe, Baton Rouge, East Baton Rouge.

SAURURACEÆ—*Lizard's Tail Family.*

Saururus cernuus, L., Lizard's Tail, near Mississippi river, West Baton Rouge.

CALLITRICHACEÆ—*Water Starwort Family.*

Callitriche pedunculosa, Nutt, Water Starwort, Baton Rouge, East Baton Rouge.

EUPHORBIACEÆ—*Spurge Family.*

Euphorbia corollata, L., Large flowering Spurge, Old Seminary, Rapides.

Euphorbia hypericifolia, L., Large spotted Spurge, Baton Rouge, East Baton Rouge.

Euphorbia hypericifolia, var. communis, Common spotted Spurge, Port Hudson, East Feliciana.

Euphorbia herniarioides, Nutt, Small-leafed Spurge, Baton Rouge, East Baton Rouge.

Euphorbia Ludoviciana, N. Sp., Louisiana Spurge, Washington, St. Landry.

Euphorbia Meganæsos, N. Sp., Grand Isle Spurge, Grand Isle, Jefferson.

Acalypha gracilens, Gray, Slender three-seeded Mercury, Old Seminary, Rapides.

Acalypha Caroliniana, Walt., Carolina, Three-seeded Mercury, Baton Rouge, East Baton Rouge.

Tragia urens, L., Stinging Tragia, Old Seminary, Rapides.

Stillingia ligustrina, Michx., Privet-leafed Stillingia, Clinton, East Feliciana.

Stillingia Sylvatica, L., Queen's Delight, Old Seminary, Rapides.

Croton Elliottii, Chapm, Elliott's Croton, Old Seminary, Rapides.

Ricinus communis, L., Castor Oil Plant, Ville Platte, St. Landry.

BATIDACEÆ—*Batis Family.*

Batis maritima, L., Seaside Batis, Grand Isle, Jefferson.

URTICACEÆ—*Nettle Family.*

Urtica gracilis, Ait., Slender-stemmed Nettle, Chicotville, St. Landry.

Urtica dioica, L., Common Nettle, Baton Rouge, East Baton Rouge.

Urtica purpurascens, Nutt, Purple Nettle, Baton Rouge, East Baton Rouge.

Parietaria debilis, Forst., Feeble-stemmed Pellitory, Baton Rouge, East Baton Rouge.

MORACEÆ—*Mulberry Family.*

Morus rubra, L., tree, Red Mulberry, Baton Rouge, East Baton Rouge.

Morus nigra, Vent., tree, Black Mulberry, native of China, introduced.

Maclura aurantiaca, Nutt, tree, Osage Orange, native of Arkansas, introduced.

Ficus carica, tree, Fig Tree, native of Asia, introduced.

ULMACEÆ—*Elm Family.*

Ulmus alata, Michx., tree, Whahoo, Baton Rouge, East Baton Rouge.

PLATANACEÆ—*Plane Tree Family.*

Platanus occidentalis, L., tree, Plane Tree, Sycamore, Washington, St. Landry.

JUGLANDACEÆ—*Walnut Family.*

Juglans nigra, L., tree, Black Walnut, Baton Rouge, East Baton Rouge.

Carya amara, Nutt, tree, Bitter Nut, Grosse Tete Railroad, West Baton Rouge.

Carya glabra, Tor., tree, Pignut, Baton Rouge, East Baton Rouge.

Carya alba, Nutt, tree, Shell-bark Hickory, Baton Rouge, East Baton Rouge.

CUPULIFERÆ—*Oak Family.*

Quercus phellos, L., tree, Willow Oak, Baton Rouge, East Baton Rouge.

Quercus aquatica, Catesby, tree, Water Oak, Old Seminary, Rapides.

Quercus aquatica, Catesty, var. hybrida, tree, Hybrid Water Oak, Baton Rouge, East Baton Rouge.

Quercus tinctoria, Bart., tree, Black Oak, Baton Rouge, East Baton Rouge.

Quercus falcata, Michx., tree, Spanish Oak, Baton Rouge, East Baton Rouge.

Quercus obtusifolia, Michx., tree, Post Oak, Baton Rouge, East Baton Rouge.

Quercus alba, L., tree, White Oak, Old Seminary, Rapides.

Quercus prinus, L., tree, Swamp Chestnut Oak, Baton Rouge, East Baton Rouge.

Quercus nigra, L., tree, Black Jack, Old Seminary, Rapides.

Quercus virens, Ait., tree, Live Oak, Brashear City, St. Mary.

Castanea pumila, Michx., tree, Chinquapin, Opelousas, St. Landry.

Fagus ferruginea, Ait., tree, American Beech, Baton Rouge, East Baton Rouge.

SALICACEÆ—*Willow Family.*

Populus angulata, Ait., tree, Cottonwood, Baton Rouge, East Baton Rouge.

Populus dilatata, tree, Lombardy Poplar, Native of Italy, Introduced.

Salix nigra, Marsh.. tree, Black Willow, Baton Rouge, East Baton Rouge.

Salix Babylonica, tree, Weeping Willow, Native of the East, Introduced.

CONIFERÆ—*Pine Family.*

Pinus Australis, Michx., tree, Long-leafed Pine, Old Seminary, Rapides.

Pinus taeda, L., tree, Old-field Pine, Old Seminary, Rapides.

Pinus glabra, Walt., tree, Smooth Pine, Amite River, East Baton Rouge.

Pinus inops, Ait., tree, Scrub Pine, Clinton, East Feliciana.

Taxodium distichum, Rich., tree, Cypress, Baton Rouge, East Baton Rouge.

Juniperus Virginianus, L., tree, Red Cedar, Baton Rouge, East Baton Rouge.

Cupressus thyoides, L., tree, White Cedar, Old Seminary, Rapides.

Thuya occidentalis, L., tree, Arbor Vitae, Native of North Caralina, Introduced.

Thuya orientalis, shrub, Oriental Arbor Vitae, Native of China, Cultivated.

Taxus Hibernica, shrub, Irish Yew, Native of Ireland, Cultivated.

MONOCOTYLEDONOUS PLANTS.

PALMÆ—*Palms.*

Sabal Adansonii, Guerns., Dwarf Palmetto, New Orleans, Orleans.

Sabal Palmetto, R. & S., Tree Palmetto, Grand Isle, Jefferson.

ARACEÆ—*Arum Family.*

Peltandra Virginica, Raf., Virginia Arrow Arum, Amite river, East Baton Rouge.

LEMNACEÆ—*Duckweed Family.*

Lemna minor, L., Small Duckweed, New Orleans, Orleans.

Lemna polyrhiza, L., Many-rooted Duckweed, New Orleans, Orleans.

Lemna Torreyi Austin, Torrey's Lemna, Washington, St. Landry.

TYPHACEÆ—*Cat Tail Family.*

Typha latifolia, L., Cat-tail, Baton Rouge, East Baton Rouge.

ALISMACEÆ—*Water Plantain Family.*

Sagittaria falcata, Pursh., Scythe-leafed Arrow-head, Lake Pontchartrain, Orleans.

Sagittaria heterophylla, Pursh., Unequal-leafed Arrow-head, Brashear City, St. Mary.

Sagittaria variabilis, Engelm., Varied-leafed Arrow-head, Franklin, St. Mary.

Sagittaria simplex, Pursh., Simple-leaved Arrow-head, Prairie lakes, St. Landry.

ORCHIDACEÆ—*Orchis Family.*

Platanthera ciliaris Lind. Orchis ciliaris, L., Fringe-flowered False Orchis, Pontchatoula, Tangipahoa.

Habenaria Michauxii, Nutt., Michaux' Orchis, Old Seminary, Rapides.

Spiranthes gracilis, Slender-twisted Orchis, Old Seminary, Rapides.

CANNACEÆ—*Canna Family*

Canna Indica, L., Indian Shot Plant, L., cultivated.

AMARYLLIDACEÆ—*Amaryllis Family.*

Hypoxis erecta, L., Erect Star Grass, Amite City, Tangipahoa.

Hypoxis Juncea, Smith, Rushy Star Grass, Grand Isle, Jefferson.

Narcissus, Tazetta, Narcissus, native of Spain, cultivated.

HÆMODORACEÆ—*Bloodwort Family.*

Aletris aurea, Walt., Golden-flowered Aletris, Old Seminary, Rapides.

BROMELIACEÆ—*Pine Apple Family.*

Tillandsia usneoides, L., Long Moss, near Mississippi river, West Baton Rouge.

IRIDACEÆ—*Iris Family.*

Iris tripetala, Walt., Three-petaled Iris, Old Seminary, Rapides.

Iris cuprea, L., Copper-flowered Iris, Baton Rouge, East Baton Rouge.

Iris Florentina, Sweet Flag, native of Italy, cultivated.

SMILACEÆ—*Smilax Family.*

Smilax tamnoides, L., vine, Green Briar, Baton Rouge, East Baton Rouge.

Smilax pseudo-China, L., vine, False Chinese Smilax, Baton Rouge, East Baton Rouge.

Trillium sessile, L., Stalkless Three-leafed Nightshade, Amite river, East Baton Rouge.

LILIACEÆ—*Lily Family.*

Lilium Lockettii, N. Sp., Lockett's Lily, Lake St. Charles, Calcasieu.

Lilium candidum, White Lily, native of Levant, cultivated.

Lilium tigrinum, Tiger-spotted Lily, native of China, cultivated.

Lilium Carolinianum, Orange-flowered Lily, Baton Rouge, East Baton Rouge.

Yucca filamentosa, L., Bear Grass, Old Seminary, Rapides.

Allium Canadense, Kalm, Wild Onion, near the Mississippi river, West Baton Rouge.

Gladiolus Byzantinus, Turkish Gladiolus, native of the Levant, cultivated.

Gladiolus Psittacinus, Corn Flag, native of South of Europe, cultivated.

MELANTHACEÆ—*Colchicum Family.*

Melanthium Virginicum, L., Bunch Flower, Amite City, Tangipahoa.

Chamælirium luteum, Gray Helonias dioica, Blazing Star.

JUNCACEÆ—*Rush Family.*

Juncus scirpoides, Lam., Scirpoid Rush, Baton Rouge, East Baton Rouge.

Juncus maritimus, Lam., Sea-side Bullrush, Baton Rouge, East Baton Rouge.

Juncus effusus, L., Soft Rush.

Juncus tenuis, Willd, Slender Rush, Baton Rouge, East Baton Rouge.

Juncus marginatus, Willd, Broad-brimmed Rush, Marsh Prairie, St. Mary.

Juncus setaceus, Rosth, Bristly Rush, Baton Rouge, East Baton Rouge.

Juncus buffonius, L., Toad Rush, Baton Rouge, East Baton Rouge.

PONTEDERIACEÆ—*Pickerel Weed Family.*

Pontederia cordata, L., Wampee, Pickerel Weed, New Orleans, Orleans.

Schollera graminea, Willd, Grass-leafed Schollera, Brashear City, St. Mary.

COMMELYNACEÆ—*Spiderwort Family.*

Commelyna communis, L., Common Dayflower, Opelousas, St. Landry.

Commelyna Virginica, Virginia Dayflower, Baton Rouge, East Baton Rouge.

Commelyna Virginica, var. angustifolia, Ell., Narrow-leafed Dayflower, New Orleans, Orleans.

Commelyna erecta, L, Erect Dayflower, Old Seminary, Rapides.

Tradescantia Virginiana, L., Virginia Spiderwort, Old Seminary, Rapides.

XYRIDACEÆ—*Yellow-eyed Grass Family.*

Xyris difformis, Chapm., Unsightly Yellow-eyed Grass, Covington, St. Tammany.

Xyris Elliotii, Chapm., Elliott's Yellow-eyed Grass, Old Seminary, Rapides.

Xyris elata, Chapm., High-stemmed Yellow-eyed Grass, Amite city, Tangipahoa.

ERIOCAULONACEÆ.

Eriocaulon decangulare, L., Pipewort, Covington, St. Tammany.

CYPRACEÆ—*Sedge Family.*

Cyperus vegetus, Willd, Stout-stemmed Cyperus.

Cyperus strigosus, L., Bristle-spiked Galingale, Ville Platte, St. Landry.

Cyperus articulatus, L., Jointed Cyperus, Franklin, St. Mary.

Cyperus rotundus, L. C., hydra, Michx, Nut Grass, Baton Rouge, East Baton Rouge.

Cyperus compressus, L., Flat-stemmed Cyperus, Franklin, St. Mary.

Cyperus flavescens, L., Yellow Cyperus, Clinton, East Feliciana.

Cyperus ovularis, Tor., Egg-spiked Cyperus, Opelousas, St. Landry.

Cyperus tetragonus, Ell, Four-angled Cyperus, Opelousas, St. Landry.

Cyperus Gatesii, Tor., Gates' Cyperus, Opelousas, St. Landry.

Cyperus virens, Michx, Evergreen Cyperus, Opelousas, St. Landry.

Cyperus speciosus, Vahl., Showy Cyperus, Bayou Portage, St. Mary.

Cyperus Haspan, L., Haspan's Cyperus, Old Seminary, Rapides.

Cyperus erythrorhizos, Muhl., Red-rooted Cyperus, Baton Rouge, East Baton Rouge.

Rhynchospera corniculata, Tor., Horned Beak Rush, Baton Rouge, East Baton Rouge.

Rhynchospera glomerata, Vahl, Bushy Rush, Baton Rouge, East Baton Rouge.

Rhynchospera cymosa, Nutt, Cyme-flowered Rush, Baton Rouge, East Baton Rouge.

Rhynchospera inexpansa, Vahl, Unexpanded Rush, Baton Rouge. East Baton Rouge.

Rhynchospera paniculata, Tor., Panicled Rush, Old Seminary, Rapides.

Eleocharis simplex, Tor., Simple Spike Rush, Baton Rouge, East Baton Rouge.

Eleocharis tuberculosa, R. Br., Tuberculated Rush, Bayou Portage, St. Mary.

Eleocharis microcarpa, Tor., Small-fruited Rush, Bayou Portage, St. Mary.

Eleocharis pygmæa, Tor., Pigmy Rush, Bayou Portage, St. Mary.

Eleocharis albida, Tor., Whitish Rush, Bayou Portage, St. Mary.

Eleocharis obtusa, R. Br., Obtuse Rush, Baton Rouge, East Baton Rouge.

Eleocharis palustris, R. Br., Swamp Rush, Baton Rouge, East Baton Rouge.

Dichromena leucocephala, Michx., White-flowered Dichromena, Grand Isle, Jefferson.

Kyllingia palustris, Michx., Swamp Kyllingia, Clinton, East Feli, ciana.

Mariscus retrofractus, Vahl, Inversely-tapering Mariscus, Baton Rouge, East Baton Rouge.

Fimbristylis spadicea, Vahl, Chestnut-brown Fimbristylis, Bayou Portage, St. Mary.

Fimbristylis autumnalis, F. & Sch., Autumnal Fimbristylis, Port Hudson, East Feliciana.

Scirpus eriophorum, Michx., Wooly Bullrush.

Scirpus palustris, L., Swamp Bullrush, Berwick's Bay, St. Mary.

Trichelostylis autumnalis, Roem. & Shultz, Autumnal Trichelostylis, Baton Rouge, East Baton Rouge.

Fuirena squarrosa, Michx., Coarse Fuirena, Baton Rouge, East Baton Rouge.

Carex lupulina, Muhl., Lupine Sedge, Baton Rouge, East Baton Rouge.

Carex glaucescens, Ell., Pale-green Sedge, Marsh Prairie, St. Mary.

Carex staminea, Willd., Staminate Sedge, Marsh Prairie, St. Mary.

Carex staminea, var. festucacea, Shrb., Fescue-grass Sedge.

Carex retroflexa, Muhl., Inverse-reflexed Sedge, Baton Rouge, East Baton Rouge.

Carex stenolepis, Tor., Narrow-scaled Sedge, Baton Rouge, East Baton Rouge.

Carex crus-corvi, Shuttleworth, Crow-legged Sedge, Cheneyville, Rapides.

GRAMINEÆ—*Grass Family.*

Leersia Virginica, Willd., Virginia Rice-grass, Old Seminary, Rapides.

Leersia Centicularis, Michx., False Rice-grass, Old Seminary, Rapides.

Agrostis perennans, Gray, Thin-Grass, Old Seminary, Rapides.

Agrostis scabra, Willd., Hair-Grass, Old Seminary, Rapides.

Cynodon Dactylon, Pers., Bermuda-Grass, Baton Rouge, East Baton Rouge.

Phleum pratense, L., Timothy, Baton Rouge, East Baton Rouge.

Poa pratensis, L., Orchard-grass, Baton Rouge, East Baton Rouge.

Poa annua, L., Low Spear-grass, Baton Rouge, East Baton Rouge.

Poa flexuosa, Muhl., Slender-stemmed Poa, Amite river, East Baton Rouge.

Aristida purpurascens, Poir., Triple-awned Aristida, Old Seminary, Rapides.

Aristida gracilis, Ell., Slender Aristida, Old Seminary, Rapides.

Aristida virgata, Trin., Strait-stemmed Aristida, Old Seminary, Rapides.

Muhlenbergia diffusa, Schreb., Drop-seed, Nimble-Will, Old Seminary, Rapides.

Eatonia Pennsylvanica, Gray, Pennsylvania Eatonia, Old Seminary, Rapides.

Eatonia obtusata, Gray, Obtuse-spiked Eatonia, Old Seminary, Rapides.

Vilfa vaginæflora, Tor., Sheath-flowered Vilfa, Cheneyville, Rapides.

Melica mutica, Walt., Beardless Melic-grass, Baton Rouge, East Baton Rouge.

Eragrostis ciliaris, Link., Fringed Eragrostis, Port Hudson, East Feliciana.

Eragrostis oxylepis, Tor., Huskpointed Eragrostis, Grand Isle, Jefferson.

Eragrostis reptans, Nees., Creeping Eragrostis, Baton Rouge, East Baton Rouge.

Eragrostis pectinacea, Gray, Comb-spiked Eragrostis, Old Seminary, Rapides.

Eragrostis capillaris, Nees, Fine-branched Eragrostis, Old Seminary, Rapides.

Eragrostis megastachya Link var. pecoides, Beauv, Large-spiked Eragrostis, Old Seminary, Rapides.

Eragrostis Purshii, Schrad, Pursh's Eragrostis, Baton Rouge, East Baton Rouge.

Eleusine Indica, Gaert, Crab grass, Yard grass, Baton Rouge, East Baton Rouge.

Paspalum digitaria, Poir, Finger-spiked Paspalum, Baton Rouge, East Baton Rouge.

Paspalum læve, Michx., Smooth Paspalum, Baton Rouge, East Baton Rouge.

Paspalum ciliatifolia, Michx., Fringe-leafed Paspalum, Old Seminary, Rapides.

Paspalum racemulosum, Nutt., Racemed Paspalum, Old Seminary, Rapides.

Paspalum praecox, Walt., Early Paspalum, near Mississippi river, West Baton Rouge.

Paspalum fluitans, Kunth, Floating Paspalum, Cheneyville, Rapides.

Sporobulus junceus, Kunth, Rush-leafed Drop-seed grass, Old Seminary, Rapides.

Sporobulus Indicus, Brown, Indian Drop-seed grass, Old Seminary, Rapides.

Setaria glauca, Beauv, Pale Green Setaria, Baton Rouge, East Baton Rouge.

Digitaria glabra, P., glabrum, Gaud., Smooth Digitaria, Baton Rouge, East Baton Rouge.

Panicum sanguinale, L., Finger grass, Crab grass, Baton Rouge, East Baton Rouge.

Panicum sanguinale, L , var. villosum, Ell, Hairy Finger grass, Baton Rouge, East Baton Rouge.

Panicum filiforme, L., Fine-stemmed Finger grass, Baton Rouge, East Baton Rouge.

Panicum dichotomum, L., var. villosum, Ell., Woolly two-branched Panic grass, Old Seminary, Rapides.

Panicum dichotomum, L., var. lanuginosum, Ell., Downy two-branched Panic grass, Old Seminary, Rapides.

Panicum dichotomum, L., var. barbulatum, Michx., Bearded two-branched Panic grass, Old Seminary, Rapides.

Panicum dichotomum, L., var. ramulosum, Michx., Branching two-branched Panic grass, Old Seminary, Rapides.

Panicum dichotomum, L., var. nitidum, Ell., Glossy two-branched Panic grass, Old Seminary, Rapides.

Panicum dichotomum, L., var. pubescens, Ell., Hairy two-branched Panic grass, Old Seminary, Rapides.

Panicum virgatus, Ell., Straight-stemmed Panic grass, Bayou Portage, St. Mary.

Panicum hians, Ell., Disjointed Panic grass, Amite River, East Baton Rouge.

Panicum gymnocarpum, Ell., Nude-fruited Panic grass, Baton Rouge, East Baton Rouge.

Panicum capillare, L., Hair-branched Panic grass, Baton Rouge, East Baton Rouge.

Panicum Walteri, Ell., Walter's Panic grass, Baton Rouge, East Baton Rouge.

Panicum agrostoides, Spreng, Bentgrass Panic grass, Old Seminary, Rapides.

Panicum hirtellum, L., Rough-stemmed Panic grass, Old Seminary, Rapides.

Panicum microcarpon, Muhl, Small-fruited Panic grass, Old Seminary, Rapides.

Panicum verucosum, Muhl, Wart-fruited Panic grass, Old Seminary, Rapides.

Panicum viscidum, Ell., Clammy Panic grass, Old Seminary, Rapides.

Panicum rufum, Kunth, Brown-haired Panic grass, Old Seminary, Rapides.

Panicum capillare, L., var. geniculatum, Muhl, Knee-jointed Panic grass, Old Seminary, Rapides.

Panicum crusgalli, L., Barnyard grass, Baton Rouge, East Baton Rouge.

Panicum crusgalli, L., var. longisetum, Long-bristled Barnyard grass, Pontchartrain, Orleans.

Uniola gracilis, Michx, Slender Uniola, Ville Platte, St. Landry.

Uniola latifolia, Michx., Broad-leafed Uniola, Old Seminary, Rapides.

Spartina juncea, Willd., Rush-leafed Marsh grass, Lake Pontchartrain, Orleans.

Chloris petræi, Stony Chloris, Grand Isle, Jefferson.

Stenotaphrum Americanum Schrank, American Stenotaphrum, Grand Isle, Jefferson.

Brizopyrum spicatum, Hook., var. macrostachyon, Large-spiked Brizopyrum, Grand Isle, Jefferson.

Coix lachryma, Lachrymal Coix, Bayou Portage, St. Mary.

Sorghum nutans, Gray, Indian grass, Wood grass, Old Seminary, Rapides.

Sorghum Halapense, Pers., Cuba grass, Lake Pontchartrain, Orleans.

Leptochloa mucronata, Kunth, Leaf-pointed Leptochloa, Baton Rouge, East Baton Rouge.

Andropogon Virginicus, L., Virginia Beard-grass, Old Seminary, Rapides.

Andropogon scoparius, Michx., Broom-grass, Old Seminary, Rapides.

Andropogon Elliottii, Chapm, Elliotti's Beard-grass, Old Seminary, Rapides.

Andropogon argenteus, Ell., Silver-spiked Beard-grass, Old Seminary, Rapides.

Glyceria aquatica, Smith, Reed Meadow grass, Old Seminary, Rapides.

Glyceria obtusa, Trin., Obtuse-spiked Meadow-grass, Old Seminary, Rapides.

Erianthus alopecuroides, Ell., Woolly Beard-grass, Old Seminary, Rapides.

Cinna arundinacea, L., Wood Reed-grass, Old Seminary, Rapides.

Festuca pratensis, Huds., Meadow Fescue grass, Baton Rouge, East Baton Rouge.

Festuca myurus, L., Knotty Fescue grass, Baton Rouge, East Baton Rouge.

Alopecurus geniculatus, L., Foxtail, Baton Rouge, East Baton Rouge.

Hordeum pratense, L., Meadow Barlow, Baton Rouge, East Baton Rouge.

Ctenium Americanum Spreng, American Toothache grass, Pontchatoula, Tangipahoa.

Cenchrus tribuloides, L., Burr-grass, Grand Isle, Jefferson.

Cenchrus echinatus, L., Hedgehog grass, Cheneyville, Rapides.

Arundinaria gigantea, Chapm, Cane, Baton Rouge, East Baton Rouge.

Avena elatior, L., High-stemmed Oats, Baton Rouge, East Baton Rouge.

Avena sativa, L., Common Oats, Baton Rouge, cultivated.

Zea Mays, Indian Corn, Baton Rouge, cultivated.

Sorghum saccharatum, Broom Corn, Baton Rouge, cultivated.

Oryza sativa, Rice, Baton Rouge, cultivated.

Saccharum officinarum, Sugar Cane, Baton Rouge, cultivated.

CRYPTOGAMOUS OR FLOWERLESS PLANTS.

Filices—*Ferns.*

Onoclea sensibilis, L., Sensitive Fern, Baton Rouge, East Baton Rouge.

Aspidium Noveboracense, Willd, Shield Fern, Old Seminary, Rapides.

Aspidium patens, Swartz, Spreading Shield Fern, Opelousas, St. Landry.

Aspidium thelypteris, Swartz, Edge-winged Fern, Baton Rouge, East Baton Rouge.

Aspidium achrostichoides, Swartz, Uncolored Fern, Old Seminary, Rapides.

Woodwardia angustifolia, Smith, Narrow-leafed Woodwardia, Old Seminary, Rapides.

Osmunda cinamomea, L., Cinnamon Fern, Old Seminary, Rapides·

Osmunda regalis, Flowering Fern, Old Seminary, Rapides.

Pteris aquilina, L., Common Brake, Old Seminary, Rapides.

Pteris caudata, L., Tail-leafed Brake, Old Seminary, Rapides.

Polypodium incanum, Swartz, Hoary Polypody, Old Seminary, Rapides.

Asplenium ebeneum, Ait., Black-stemmed Spleenwort, Baton Rouge, East Baton Rouge.

Asplenium ebeneum, Ait., var. Bacculum Rubrum, Baton Rouge Spleenwort, Baton Rouge, East Baton Rouge.

Asplenium filix foemina, R. B., Female Fern Spleenwort, Old Seminary, Rapides.

LYCOPODIACEÆ—*Club Moss Family.*

Selaginella apus, Spring, Martinet Selaginella, Baton Rouge, East Baton Rouge.

HYDROPTERIDES—*Water Fern Family.*

Azolla Caroliniana, Willd, Carolina Azolla, Port Hudson, East Feliciana.

MUSCI—*Mosses.*

Thuidium gracile, Br. and Sch., Slender Thuidium, Baton Rouge, East Baton Rouge.

Thuidium tamariscinum, Hedw. Hypnum, Tamarik-leafed Thuidium, Baton Rouge, East Baton Rouge.

Amblystegium radicale, Brid. Hypnum, Root-leafed Amblystegium, Baton Rouge, East Baton Rouge.

Cylindrothecium seductrix, Hedw., Fair-leafed Cylindrothecium, Baton Rouge, East Baton Rouge.

Cylindrothecium cladœrhizans, Hedw., Branch-rooting Cylindrothecium, Baton Rouge, East Baton Rouge.

Atrichum angustatum, Hook, Narrow-leafed Shield Moss, Baton Rouge, East Baton Rouge.

Atrichum crispum, James, Crisp-leafed Shield Moss, Baton Rouge, East Baton Rouge.

Weissia viridula, Brid., Green Pearl Moss, Baton Rouge, East Baton Rouge.

Trichostomum pallidum, Hedw., Pale Trichostomum, Baton Rouge, East Baton Rouge.

Physcomitrium pyriforme, L., Pear-fruited Physcomitrium, Baton Rouge, East Baton Rouge.

Funaria flaviscus, Michx, Yellow Funaria, Baton Rouge, East Baton Rouge.

Funaria hygrometrica, Hedw., Common Funaria, Baton Rouge, East Baton Rouge.

Desmatodon plinthobius, Sul. & Lesq., Plinth-fruited Desmatodon, Baton Rouge, East Baton Rouge.

Bartramia radicalis, Beauv., Root-leafed Bartramia, Baton Rouge, East Baton Rouge.

Eurhynchia hians, Hedw., Disjointed Eurhynchia, Baton Rouge, East Baton Rouge.

Leucobryum glaucum, Pale-green White Moss, Old Seminary, Rapides.

Leucobryum minus, Hampe, Small-leafed White Moss, Old Seminary, Rapides.

Rhynchostegium microcarpum, T. Mull., Small-fruited Rhynchostegium, Baton Rouge, East Baton Rouge.

Leptodon trichomittium, Mohr. var. immorsus, Sul., Hair-capped Leptodon, Baton Rouge, East Baton Rouge.

Hypnum curvifolium, Hedw., Curve-leafed Branch Moss, Baton Rouge, East Baton Rouge.

Hypnum micans, Schwartz, Shining Branch Moss, Baton Rouge, East Baton Rouge.

Hypnum micans var. albulum, Aust, Whitish Branch Moss, Baton Rouge, East Baton Rouge.

Hypnum recurvans, Br., Bent Branch Moss, Baton Rouge, East Baton Rouge.

Sphagnum cymbifolium, Dill, Boat-leafed Sphagnum, Old Seminary, Rapides.

Leskea polycarpa, Hedw., Many-fruited Leskea, Baton Rouge, East Baton Rouge.

Leucodon julaceus, Hedw., Close-matted Leucodon, Baton Rouge, East Baton Rouge.

Fissidens adiantoides, Hedw., Maiden-hair Fissidens, Baton Rouge, East Baton Rouge.

Fissidens minutulus Sul var terrestris, Diminutive Fisssidens, Baton Rouge, East Baton Rouge.

Barbula muralis, Hedw., Wall Screw moss, Baton Rouge, East Baton Rouge.

Barbula sabciliata, Aust., n. sp., Sparse-Fringed Beard moss, Baton Rouge, East Baton Rouge.

Brachythecium rutabulum, Br., Rake-leafed Brachythecium, Baton Rouge, East Baton Rouge.

Brachythecium plumosum, Br., Feather-leafed Brachythecum, Baton Rouge, East Baton Rouge.

Meteorium pendulum, Sull., Festoon-branched Meteorium, Baton Rouge, East Baton Rouge.

Cryphea nervosa, Hook and Wils., Nerve-leafed Cryphea, Baton Rouge, East Baton Rouge.

Cryphea glomerata, Sch., Tufted Cryphea, Baton Rouge, East Baton Rouge.

Mnium affine, Bland, Close-marked Star moss, Baton Rouge, East Baton Rouge.

Trematodon longicollis, Michx, Long-necked Trematodon, Baton Rouge, East Baton Rouge.

Hepaticæ—*Liverwort Family.*

Lejeunia clypeata, Sull, Shield-leafed Lejeunia, Baton Rouge, East Baton Rouge.

Lejeunia Sullivantii, Aust., Sullivant's Lejeunia, Baton Rouge, East Baton Rouge.

Lophocolea heterophylla, Nees, Mixed-leafed Lophocolea, Baton Rouge, East Baton Rouge.

Plagiochyla Ludoviciana, Sul., Louisiana Plagiochyla, Baton Rouge, East Baton Rouge.

Jungermannia Schraderi, Mort, Schrader's Jungermannia, Baton Rouge, East Baton Rouge.

Anthoceros lævis, L., Smooth Anthoceros, Old Seminary, Rapides.

Dumortiera hirsuta, Nees, Rough-leafed Dumortiera, Amite river, East Baton Rouge.

Steezia Lyellii, Lehm., Lyell's Steezia, Old Seminary, Rapides.

Fimbriana tenelle, Nees, Slender Fimbriana, Baton Rouge, East Baton Rouge.

Marchantia polymorpha, L., Multiform Marchantia, Baton Rouge, East Baton Rouge.

Madotheca Porella, Dicks., Porella's Madotheca, Baton Rouge, East Baton Rouge.

Madotheca involuta, Hampe, Covered Madotheca, Baton Rouge, East Baton Rouge.

Radula Lescurii, Aust., Lescure's Radula, Covington, St. Tammany.

Sphagnoecetis communis, Nees., Common Sphagnoecetis, Covington, St. Tammany.

Riccia natans, L., Floating Riccia, Baton Rouge, East Baton Rouge.

Riccia glauca, (?) L., Pale Green Riccia, Baton Rouge, East Baton Rouge.

LICHENES—*Lichens.*

Cladonia Floerkeana, Fries., Floerke's Cladonia, Amite River, East Baton Rouge.

Cladonia cristella, Tuckerm., Tufted Cladonia, Baton Rouge, East Baton Rouge.

Cladonia mitrula, Tuckerm., Mitreform Cladonia, Amite River, East Baton Rouge.

Cladonia fimbriata, Fr., var. aspersa, Tuck., Wide-spread Cladonia, Baton Rouge, East Baton Rouge.

Biatoria decolorans, Fries., Discolored Biatoria, Old Seminary, Rapides.

Biatoria russula, Tuckerm., Flesh-colored Biatoria, Baton Rouge, East Baton Rouge.

Physcia stellaris, Wollr., var. tribacia, Mean Starry Physcia, Baton Rouge, East Baton Rouge.

Physcia stellaris, Wollr., var. Dorningensis, Domingo Starry Physcia, Baton Rouge, East Baton Rouge.

Physcia picta, Nyl., Pictured Physcia, Baton Rouge, East Baton Rouge.

Physcia speciosa, var. galactopyhlla, Tuck., White-leafed Showy Physcia, Baton Rouge, East Baton Rouge.

Pyxine cocoes, var. sorediata, Nyl., Sorediate Pyxine, Baton Rouge, East Baton Rouge.

Theloschistes chrysophthalmus, var. flavicans, Yellow Theloschistes, Baton Rouge, East Baton Rouge.

Ramalina gracilenta, Ach., Slender-branched Ramalina, Baton Rouge, East Baton Rouge.

Ramalina Montagnei, De Not., Montagne's Ramalina, Baton Rouge, East Baton Rouge.

Ramalina calicaris, var. fraxinea, Ach., Ash Ramalina, Baton Rouge, East Baton Rouge.

Usnea barbata, Fr., var. florida, Fr., Flowering Bearded Usnea, Red River Falls, Rapides.

Usnea trichodea, Ach., Hair-branched Usnea, Red River Falls, Rapides.

Graphis scripta, Ach., Written-marked Graphis, Baton Rouge, East Baton Rouge.

Graphis sculpturata, Carved-marked Graphis, Baton Rouge, East Baton Rouge.

Lecanora pallescens, Fr., Pale Lecanora, Baton Rouge, East Baton Rouge.

Lecanora subfusca, Ach., Brownish Lecanora, Baton Rouge, East Baton Rouge.

Trypethelicum cruentum, Mont., Red Trypethelicum, Baton Rouge, East Baton Rouge.

Buellia parasema, Koerb, False Buellia, Baton Rouge, East Baton Rouge.

Heterothecium Domingense, Tuck., Domingo Heterothecium, Baton Rouge, East Baton Rouge.

Pyrenula nitida, Ach., Glossy Pyrenula, Baton Rouge, East Baton Rouge.

Pertussaria velata, Nyl, Covered Pertussaria, Baton Rouge, East Baton Rouge.

Pertussaria leioplaca, Ach., Many scaled Pertussaria, Baton Rouge, East Baton Rouge.

Parmelia perforata, Ach., Perforated Parmelia, Baton Rouge, East Baton Rouge.

FUNGI.

Agaricus campestris, L., Field Mushroom, Baton Rouge.

Agaricus amygdalinus, Curt., Almond Mushroom, Baton Rouge, East Baton Rouge.

Agaricus cespitosum, Curt., Tufted Mushroom, Baton Rouge, East Baton Rouge.

Polyporus laceratus, Berk., Ragged Polyporus, Baton Rouge, East Baton Rouge.

Polyporus sanguineus, Fr., Blood-red Polyporus, Baton Rouge, East Baton Rouge.

Polyporus lucidus, Fr., Shining Polyporus, Baton Rouge, East Baton Rouge.

Schizophyllum commune, Fr., Common Schizophyllum, Baton Rouge, East Baton Rouge.

Lentinus Lecontei, Fr., Leconte's Lentinus, Baton Rouge, East Baton Rouge.

Lentinus Ravenellii, Berk. & Curt., Ravenel's Lentinus, Baton Rouge, East Raton Rouge.

Lentinus lepideus, Fr., Smooth Lentinus, Baton Rouge, East Baton Rouge.

Atractium coccigena, Berk. & Curt., Scarlet Atractium, Baton Rouge, East Baton Rouge.

Stilbum cinnabarinum, Mont., Cinnabar-colored Stilbum, Baton Rouge, East Baton Rouge.

Melogramma gyrosum, Schw., Winding Melogramma, Baton Rouge, East Baton Rouge.

Merulius brassicæfolius, Schw., Cabbage-leafed Merulius, Baton Rouge, East Baton Rouge.

Aethalium septicum, Fr., Putrid Aethalium, Baton Rouge, East Baton Rouge.

Diatrype platystoma, Schw., Wide-mouthed Diatrype, Baton Rouge, East Baton Rouge.

Uredo Potentillarum, D. C., Cinqfoil Uredo, Baton Rouge, East Baton Rouge.

Uredo luminata, Schw., Blabkberry Uredo, Baton Rouge, East Baton Rouge.

Erineum fagineum, Pers., Beech Erineum, Baton Rouge, East Baton Rouge.

Hydnum ochraceum, Pers., Yellow Hydnum, Baton Rouge, East Baton Rouge.

Irpex bicolor, Berk. & Curt., Two-colored Irpex, Baton Rouge, East Baton Rouge.

A number of the Fungi have not yet been determined.

ALGÆ.

OSCILLARIACEÆ.

Vibrio Bacillus, Müll., West Baton Rouge.

Oscillaria—several species undetermined—Baton Rouge.

NOSTOCACEÆ.

Anabaena circinalis (?), Rab., West Baton Rouge.

SCYTONOMACEÆ.

Calothrix cæspitora (?), K., Baton Rouge.

Tolypothrix distorta (?), New Orleans.

DESMIDIEÆ.

Closterium acerosum (?), E., West Baton Rouge.

Closterium moniliforme (?), E., West Baton Rouge.

Closterium lanceolatum (?), K., Baton Rouge.

Closterium Leibleinii, K., Baton Rouge.

Closterium striolatum, Ehr., Baton Rouge.

Docidium Bacculum, Breb., Rapides.

Euastrum didelta, Ralfs., Rapides.

ZYGNEMACEÆ.

Spirogyra quinina, Ag.

Spirogyra nitida, Dillw.

Spirogyra decimina (?), Müller.

Zygnema cruciata, Ag.

NEMATOPHYCEÆ.

Bulbochaete setigera, Ag., Lake Pontchartrain.

Schizomeris Leibleinii (?), K., Grand Isle.

Drapnaldia glomerata, Vauch., Baton Rouge.

There are a great number of specimens of the Confervoid Algæ and of the Ulvaceæ and the genus Closterium undetermined, and many given in the list are probably only approximations.

DIATOMACEÆ.

EUNOTIÆ.

Epithemia musculus, K., Lake Pontchartrain.

Epithemia turgida (?), K., Lake Pontchartrain.

Epithemia gibba, K., Baton Rouge.

Epithemia Westermannii, K., Baton Rouge.

Eunotia tetraodon Ehr., Rapides.

Himantidium pectinale, K., in long filaments and separate valves, Amite river.

Himantidium bidens, E., Rapides.

Himantidium, Arcus, E., Rapides.

Himantidium undulatum, S., Rapides. This species has a gibbous center, three slight dorsal undulations, and obtuse recurved apices, and corresponds with the description of it in Pritchard. The habitat indicated in that work is Europe.

Himantidium parallelum, E.

FRAGILLARIACEÆ.

Fragillaria capucina, Desm., in elongated tapering filaments, Baton Rouge.

Nitzschia parvula, S., Grand Isle.

Nitzschia Dianæ, E., Pontchartrain.

Nitzschia amphioxys, E., Baton Rouge.

Ceratoneis Closterium, E., Rapides.

SURIRELLEÆ.

Synedra vitrea, K, Baton Rouge. The valve of this species is beautifully striated, with a quadrangular vacant space in the center. It is linear lanceolate, contracted to slightly dilated, obtuse, rounded ends. The front view is also perfectly transparent; very long; has distinct marginal striæ, and more or less dilated truncate ends. I have named it S. vitræ, with the charateristics of which I am only acquainted from the short description in Pritchard, because Professor Bailey gives the S. vitræ as the only species found in Southern waters. Otherwise I might have designated it by S. lanecolata.

Synedra amphirhynchus (?) E., Amite river. Only distinguished from the preceding in having the front view perfectly linear, with truncate but not dilated ends. The form of the valves is nearly the same as that of the preceding species, but it has no quadrangular median space. It is probable that neither of the above designations of Synedra are the proper ones. My object was to include them in the list by some specific name, the character of which they approximate.

Synedra lunaris, E., Amite river. A great number of Synedras are still undetermined.

Cymatopleura elliptica (?), Breb., Grand Isle. We have but a single specimen in side view.

Surirella splendida, Ehr., (we have specimens of the large and stunted variety) Rapides and Baton Rouge.

Surirella Craticula, Ehr., Lake Pontchartrain.

Surirella striatula, Turp., Lake Pontchartrain.

Surirella ovata (?) Ehr., Baton Rouge.

Surirella biseriata, Breb., Rapides.

Striatellieæ.

Tabellaria fenestrata, K., Rapides.

Melosireæ.

Melosira nummuloides, Ag., Grand Isle.

Melosira Jurgensii (?), Ag., Grand Isle.

Melosira varians (?), Ag., (In filaments of great length) Baton Rouge.

Pixidicula compressa, Bail, Grand Isle. This is one of the most beautiful diatoms of the collection. Its dotted lines reflect the light in a variety of colors.

Coscinodisceæ.

Coscinodiscus radiatus? Ehr., (many specimens,) Grand Isle.

The specimens are comparatively small, the punctæ are cellulose, smaller in the margin, and the radiated lines are obscure; but the punctæ are all distinct and very large.

Coscinodiscus subtilis? Ehr. Punctiform in radiating triangular linear rays. Grand Isle.

Actinoptychus undulatus, Ehr., (one single specimen,) Grand Isle.

BIDULPHIEÆ.

Bidulphia obtusa, Rich, (only one specimen,) Grand Isle. [Corresponds precisely with the description and figure of Pritchard.]

Odentella polymorpha, Rab., (A great number of specimens,) Grand Isle.

TERPSINOEÆ.

Terpsinoe musica, Ehr., (A great number of specimens,) Grand Isle.

COCCONEIDEÆ.

Cocconeis Pediculus? E., (Found collected in great numbers,) Lake Pontchartrain.

This species resembles C. placentula, but it is smaller, and has no transverse striæ. The longitudinal striæ are undulating, from three to four on each side of the median line. In most of the specimens the striæ are obscure. They are congregated in large numbers, on Enteromorpha.

Cocconeis Finnica, E., (This is a very beautiful species of Cocconeis,) Lake Pontchartrain.

This Cocconeis corresponds with Ehrenberg's figure; it has transverse striæ, and a lanceolate longitudinal fascia bisected by median line and nodule.

We have many other specimens of Cocconeis, but they are undetermined.

ACHNANTHEÆ.

Achnanthes subsessilis (?), K.—(numerous specimens)—Grand Isle. Some of these specimens are very beautiful.

CYMBELEÆ.

Cymbella Ehrenbergii, K. (—numerous specimens)—Baton Rouge. There are two varieties of this species, one broad lanceolate in navicular form, and the other small and rounded, with slightly produced rostrate apices. Both varieties are distinctly striated.

Cymbella maculata, K., Baton Rouge. There are several other species of cymbellæ not determined.

Amphora lybica, E., Lake Pontchartrain and Amite river.

Amphora lineolata, E., Lake Pontchartrain.

GOMPHONEMEÆ.

Gomphonema coronatum, E.—(numerous specimens) — Amite river. This is a very graceful and beautiful species.

Gomphonema Turris, E., Amite river.

Gomphonema dichotomum, K., Baton Rouge.

Gomphonema glans, E., Baton Rouge.

Gomphonema constrictum (?), E. Amite river.

NAVICULEÆ.

Navicula leptogongyla (?), E., Amite river.

Navicula viridis, Nitzsch., Amite river.

Navicula major (?), K., Amite river. This specimen is one of the finest in the collection. Corresponds with Pritchard's figure Pl. VII., but has no tumid centre.

Navicula mesogongyla (?), E., Rapides.

Navicula affinis, E., Rapides.

Navicula amphigomphus, E., Rapides and Amite River.

Navicula amphisphenia, (?) E., (specimens in great number,) West Baton Rouge. This species resembles N. cuspidata, but is somewhat ventricose, or rather rhombic lanceolate in outline. It has the nodule elongated. It is generally perfectly smooth, but sometimes a specimen is found which is closely and most finely striated. Striae are perfectly perpendicular and parallel and reach the median line. There are numerous specimens in front, side and oblique view. When alive they are yellow and seem to be covered by a membrane, having a linear, broad, transverse band in the center. Some of the specimens are very large, measuring over an inch in size, others are comparatively small. The valve is subacute, but any modification of its lateral position renders the apices broader in proportion as it approaches the front view, which is nearly linear.

Navicula Fusidium, E., Rapides.

Navicula Silicula, E., Baton Rouge.

Navicula hemiptera, (?) E., Pontchartrain. This specimen is finely striated.

Navicula crabro, E., Pontchartrain.

Navicula didyma, E., Pontchartrain.

Navicula dicephala, E., Rapides.

Navicula tabellaria, E., Rapides.

Navicula cuspidata, K., Rapides.

Navicula amphirhynchus, Baton Rouge.

Navicula rhyncocephala (?) Fr. Differs from the figure in Pritchard by tapering more gradually to the rostrate apices, and striæ only partially visible.

There are a great number of Naviculæ undetermined.

Stauroneis Phœnicentron, E., Rapides.

Stauroneis platystoma (?), E., Amite river.

Stauroneis semen, E., Lake Pontchartrain.

There are many fine specimens of Stauroneis undetermined.

Pleurosigma obtusatum, Sulliv, Baton Rouge.

Pleurosigma attenuatum, K., Lake Pontchartrain.

Pleurosigma curvulum, E., Baton Rouge.

Algæ Marinæ—*Fucaceæ.*

Sargassum vulgare, Aq., Beach of Grand Isle.

The Algæ proper, called "Seaweeds," are rare on the Louisiana Gulf coast, for there are no rocks to which they can attach themselves. We have several specimens of the class of Confervæ and Ulvaceæ, but they are not determined.

LIST OF INFUSORIA.

Want of time has prevented me from paying much attention to the Infusoriæ. The following specimens, among many others observed, have been marked:

Gonium tranquillum, Pritch.
Rotifer vulgare, Pritch.
Euglena viridis, Pritch.
Euglena sanguinea, Pritch.
Kerona polyporum (?) Pritch.
Euglypha tuberculata, Pritch.
Lepadella emarginata, Pritchard.

LIST OF SHELLS COLLECTED

Which have been Determined.

Spirula Peronii, Grand Isle.
Donax variabilis, Grand Isle.
Tellina Iris, Grand Isle.
Tellina Cayennensis, Grand Isle.
Littorina innovata, Grand Isle.
Purpura Floridana, Grand Isle.
Natica duplicata, Grand Isle.
Busycon pyrum, Grand Isle.
Modiola plicatula, Grand Isle.
Solecurtus Caribæus, Lam., Grand Isle.
Arca Campeachensis, Grand Isle.
Arca incongrua, Grand Isle.
Cardium muricatum, ballast heap, New Orleans.
Plicatula ramosa, ballast heap, New Orleans.
Chama macerophylla, Chemn., ballast heap, New Orleans,
Venus cancellata, ballast heap, New Orleans.
Helix aspersa, Müller, Baton Rouge.
Succinea ovalis, Gould, Baton Rouge.
Limnæa umbilicata, Bin., Baton Rouge.
Gnathodon cuneatus, Gray, Lake Pontchartrain.
Glandina bullata, Baton Rouge.
Melantho decisa, Say, Washington, St. Landry.
Helix auriforma, Say, Baton Rouge.
Helix thyroides Say, Baton Rouge.

The Unios collected on the banks of the Teche, and a number of shells belonging to the genus Ostrea, and some other small marine shells, I have not yet been able to have authentically identified.

All of which is respectfully submitted,

AMERICUS FEATHERMAN,
Lecturer on Botany and Professor of Modern Languages.

Respectfully forwarded to the Honorable Board of Supervisors.

D. F. BOYD,
Superintendent.

www.ingramcontent.com/pod-product-compliance
Lightning Source LLC
LaVergne TN
LVHW021417110826
845150LV00007B/1968

9781425509446